THE INNOVATIVE SOCIETY

Five Essential Conditions for Fostering Innovative Activity

Rudolph M. Kalafus

INFINITY PUBLISHING

ISBN 978-0-7414-9765-9 Paperback
ISBN 978-0-7414-9766-6 Hardcover
ISBN 978-0-7414-9767-3 Audiobook
ISBN 978-0-7414-9768-0 eBook
Library of Congress Control Number: 2013912938

Printed in the United States of America

Published January 2014

INFINITY PUBLISHING
1094 New DeHaven Street, Suite 100
West Conshohocken, PA 19428-2713
Toll-free (877) BUY BOOK
Local Phone (610) 941-9999
Fax (610) 941-9959
Info@buybooksontheweb.com
www.buybooksontheweb.com

This book is dedicated to my wife, Lois.

Contents

1. Introduction

Innovation—the word conjures up images of bright, enthusiastic people excitedly discussing ideas that will make them rich or famous and of television ads that portray companies where innovation is on every employee's lips. Governments and universities talk about it endlessly—how can we encourage it and get people to think outside the box? How can we make it part of the corporate modus operandi? Western societies have come to consider it as virtually synonymous with progress and prosperity. Innovation is seen as an essential part of modern life.

One of the main reasons I wrote this book is that when I read historians' works on technology, creativity, and economics, I found their discussions of the causes and conditions for innovative activity to be unsatisfying. As an engineer working with highly talented, innovative people, I felt these discussions seemed remote from what I was seeing on a daily basis. I wanted to clearly understand the conditions in a society that supported and encouraged innovative activity. If they were properly identified, then it should be possible to look back in history and explain why some societies were more innovative than others. One should then be able to apply these conditions to today's societies and institutions and formulate suggestions for encouraging innovative thinking at every level: in government, in schools and universities, in companies, and even in everyday life. In my opinion, previous work did not meet these criteria.

I believe that my experience in technology research and development is an asset in addressing these issues. This experience required distinguishing between "necessary" and "sufficient" conditions; between hypotheses, correlations, and theories; and between rules and guidelines that are generally valid and those that apply only under certain conditions. These distinctions are important in determining the factors that affect innovative activity. I also believe that being born into a blue-collar environment, being educated in an excellent school system, and attending a major university give me a broad-based perspective from which to view the subject: I grew up around tinkerers, spent my university years around theoreticians, and then worked with inventors and entrepreneurs.

Because I'm an engineer, I have a particular interest in technical innovations, but the innovative spirit affects other aspects of our lives. Societies that consciously encourage innovation in science and technology usually encourage similar creativity in all modes of thought and activity because the associated societal attitudes encourage questioning current practices and improving upon them. In such an environment, people can question the conventional wisdom in politics, religion, and academic subjects.

To gain an understanding of the subject, I found it useful to begin with the innovations of the ancient Egyptians, Mesopotamians, Assyrians, Hebrews, Chinese, Indians, Greeks, and Romans. They developed practical technologies such as stonecutting, sailing, aqueducts, irrigation methods, glassblowing, and metalworking. They developed legal, governmental, and philosophical theories. They developed the roots of modern mathematics and science. The Greeks brought us a whole new way of thinking. The Romans brought us new ways of organizing government. Creative and innovative people have been active for a long time.

For the first few centuries AD, innovative activity occurred primarily in Asia, rather than in Europe and around the Mediterranean. Then Islamic society encouraged curious scholars to harvest the wisdom of the earlier ages, and for about five hundred years, that society produced innovations in medicine, physics, engineering, philosophy, and mathematics. Europe also began to stir. In early medieval times, practical innovations were introduced by people whose names are long forgotten, but whose efforts had significant consequences later. The Catholic Church's iron grip on philosophical thought eventually loosened, the Protestant Reformation brought competing religious doctrines into being, contact with foreign cultures accelerated, and a fundamental shift in people's thinking took place. During the Enlightenment, this shift in thinking introduced the concepts that reason was accessible to everyone, that change was not always threatening, and that progress through change was a real possibility. Innovation came to be valued in its own right, something to be embraced rather than feared. With increased trade came mercantilism and the Industrial Revolution. Innovations were recognized as not only intellectually interesting but also economically beneficial. If new inventions could bring prosperity, changes were possible in all walks of life. In this spirit, democracy in America was born. Finally, the last two centuries have brought an explosion of innovation that can only be described as breathtaking. It is a remarkable story.

The word "innovation" has a specific meaning in this book. *Webster's Third New International Dictionary* defines innovation as "the introduction of something new" or "something that deviates from established doctrine or practice." The *Oxford English Dictionary* defines it as "making changes to something established." These definitions form a large umbrella—innovations would not be limited to science and technology. Within this broad definition, innovation takes place not only in sciences, medicine, and engineering but also in literature, law, art, philosophy, politics, education, or any subject of endeavor.

However, this book focuses on more practical types of innovation; many of the concepts discussed here are not as directly applicable to other creative endeavors like the arts and scholarship. Namely, *a device, procedure, concept, theory, or organizational structure is considered to be innovative if it is identifiable as new and is either useful or descriptive of the physical world.* This definition refers to ideas and inventions that are novel and that, at least in some small way, change the way a society views the physical world or how it functions.

Devices and inventions are what most people first picture when they hear the word "innovation." They think of computers, cars, telephones, and other practical machines, but the word also describes less tangible improvements. It can also involve improving a procedure; for example, double-entry bookkeeping was a major advance in commercial activities that enabled businessmen to have a more complete knowledge of their businesses. Concepts of the physical world, while not immediately practical, often involved huge leaps of intellect and imagination to describe the world in new ways. Theories such as that of James Clerk Maxwell connected disparate physical phenomena in novel and useful ways; he showed that radio, infrared, and light waves are essentially the same phenomenon (i.e., electromagnetic radiation) but possessing different frequencies. New organizational structures like limited liability companies have had a huge impact on the economies of the world. These are the types of innovations addressed in this book.

Very few innovations coming from tinkerers and individual inventors get widely adopted in their original forms. This is true at every level of sophistication. In Thomas Edison's time, out of hundreds of patents taken out on recording techniques, only a few withstood the test of time. Like seeds tossed on the top of the ground, only some take root and provide the foundation of major innovations. Some authors claim that the tinkerers' efforts should not be considered truly innovative, that such a noble term should be reserved for impressive works of the mind. I reject such a limiting concept, as most inventions were the product of a long chain of incremental improvements involving numerous trials and experiments, as well as the inspiration, intelligence, and perseverance of many smart people.

The concept of innovation as something desirable is a relatively recent phenomenon. Because it questioned the existing dogma, innovation was often considered to be dangerous. Even today, some cultures are uncomfortable with it. To understand why, it is necessary to understand the forces and attitudes in societies that encourage innovation and those that discourage it. For example, many of us would like to believe that tolerance and public education play a critical role, but is this really borne out by events? What about war, peace, and religion? Are there lessons from the past that can be applied in today's world?

I am not a historian, so I have drawn from material written by professional historians and writers. I have no new historical discoveries to share, but I believe my technical training gives me the tools to assess the relative importance of different factors and conditions that have been identified by historians. I believe that the focus on innovation in this book offers a perspective distinctly different from previous accounts.

In particular, I make a distinction between "blue-collar" and "white-collar" innovative activity. This distinction is very useful in describing the different sources of the creative energies at work, right up to about three centuries ago.

Blue-collar innovations refer to tangible inventions, usually produced by working people. Historically, these inventors were typically illiterate and were attempting to make a specific improvement of some sort to meet a known need, or they were tinkerers who experimented to see what would happen if they made changes to existing processes: an ironworker who pounded out a different shape for a plow to make it move more easily through the soil, a bronze producer who experimented with different amounts of tin and copper, a chemist who mixed different substances and observed the properties of the resulting compounds, a glassblower who introduced substances into the melted glass to produce new colors. These people were what I call blue-collar innovators. They were usually born into the lower classes, and they performed the practical tasks of the societies. Their innovations were usually passed down through pictures and oral means from one generation to the next.

White-collar innovations refer to less tangible inventions, such as mathematical techniques, useful procedures, and observations of patterns in nature: the mathematicians who recognized the usefulness of the Pythagorean theorem in construction, the scientist who deduced that meteorites fall to the earth from space, the scientists who modeled the solar system as a sun with planets orbiting around it. These were what I call white-collar innovative activities. They were typically conducted by people born into the privileged classes, who served in the courts of the emperors as advisers, scribes, and priests. They usually disdained the blue-collar innovative activities and the people who took part in them. Their innovations were often passed down from one generation to the next using the written word, as well as through mentoring and training. There was some crossover between the two types of innovation: since members of the privileged classes were in charge of managing the economies, some of them learned about blue-collar techniques and methods and incorporated them into the written societal knowledge base.

Once literacy became widespread, many more blue-collar innovators had access to information and training. This meant that more complex information could be passed on via the written word and diagrams, and blue-collar ideas morphed into technology. At the same time, white-collar innovators began to value experiments and observations as much as abstract thinking and began to get their hands dirty, and science as it is known today was born. Once both kinds of innovative activity were exploring the same natural universe, it was only a matter of time until the worlds met.

During the explosion of innovation of the last few hundred years, many inventions involved a combination of scientific concepts and technological knowledge. The quest to harness steam to drain water from mines and to move large objects culminated in the invention of the steam engine, after a century of efforts that combined the two kinds of knowledge. It was one of the earliest and most powerful examples of the marriage of science and technology.

Scientific advances in atomic and molecular theory have brought us knowledge of the atomic structure of materials, which have enabled us to understand chemistry and the structure of materials at a significantly deeper level. For example, semiconductors have quantum mechanical properties that make the transistor possible. Diodes, triodes, and amplifiers were transistorized, and further innovations reduced the size of these circuits to unbelievably small sizes so that today millions of these are routinely etched onto silicon wafers a few millimeters wide. To achieve this required a huge amount of theoretical and practical technological knowledge that grew rapidly and involved innovation in optics, etching techniques, air-cleaning techniques, materials, and many other fields.

There are still examples today of pure blue-collar innovative activities, as can be seen in magazines such as *Popular Mechanics*. Other examples can be seen at art fairs, where some artists seek out new materials to work with. I still come across novel designs for bread knives and new hand tools. One can find these and many other clever things on the internet. White-collar innovation continues unabated in mathematics and the sciences, and now new ideas come from all over the world, reflecting the surges in literacy and access to education in developing countries. The public fascination for new things has increased general respect for creativity of all kinds.

Another different perspective that I bring to the table is the importance I attach to the form of government in a society—in particular, the extent to which power is concentrated in the hands of a few. While innovation occurred in autocracies, it was severely circumscribed. I will also show that it only rarely occurred for the benefit of the populace; the people in power discouraged such activity. I will also show that some types of innovation have a democratizing effect (i.e., that they empower ordinary people).

When I began researching this subject, I thought I might find that innovations historically set the stage for cultural and political change, but I found that I couldn't make a very convincing case. In fact, the opposite appears to be true—namely, political and cultural changes initially set the stage for the rapid growth in innovative activity. Once that happened, however, certain key inventions and essential types of innovations did indeed spur the development of democratic institutions, which in turn encouraged innovation, creating a virtuous circle. In this book I will describe in some detail how this came about.

Reviewing historical material has given me a renewed appreciation for how difficult it is to be truly objective, and perhaps it is ultimately impossible. I can see that even the best historians, attempting to be as objective and intellectually honest as they can, have biases that show through their work, biases that become more obvious as time passes. There are also a number of historians with stated agendas, who assert, usually with good reason, that previous writers have ignored or even denied the contributions of minorities, women, lower-class people, or specific cultures, and they focus on those contributions to try to help balance the historical record. It seems especially difficult for anyone to assess the contributions of different cultures because there is a strong urge within all of us to consider our own culture to be the best. For example, Islamic philosophers and scientists have been characterized on the one hand by European scholars as merely parroting earlier thinkers or only preserving knowledge that would later be utilized; on the other hand, they have been described as brilliant scholars and thinkers in their own right who made significant original contributions. As another example, some scholars have derided inventors among working people as "low mechanicks," and their discoveries and innovations have been dismissed as lucky accidents. Biases are clearly at work.

It's not just our natural arrogance that we need to confront. We must also confront our tendency to think of our own situation as more typical than it really is in other parts of the world and than it really was in history. We have to recognize that people did think of their world differently than we do today. While our brains are constructed similarly at birth, our experiences significantly affect our opinions, our assumptions, and even the way we think. Our everyday vocabulary is different—for example, the words "science," "technology," "innovation," and "creativity," as we think of them today, did not exist in any language until recently. To inquire about the attitude of a pharaoh toward innovative activity would have been meaningless. Studies and observations of nature were considered philosophical, not scientific. People didn't appreciate the difference between speed and acceleration, and it took a long time before the concept of instantaneous motion was understood. Today we take it for granted that we can travel comfortably in a car or airplane at high speeds and say that we were going at sixty or six hundred miles per hour at some point in time. We also understand that we get pushed into the seat only when we accelerate in a car or take off in an airplane and pulled away from it when we brake or slow down. As difficult as it is, we need to attempt to see the world from the points of view of the people of earlier times.

We've all heard the saying that necessity is the mother of invention, but it would be more accurate to say that necessity is the father of invention. It is the father who traditionally tends to the needs of the family, and who is more likely to look for ways to improve his ability to provide them. At the societal level this translates into agricultural, construction, and military inventions, where the needs of food, shelter and security dominate. On the other hand the mother of invention is desire—the desire to add beauty and variety to life to make it more enjoyable. It is the mother who finds new ways of preparing food and adding spices and sauces to make it more interesting, and who finds ways of making clothing more attractive, with different colors, weaves, jewelry, and designs. It is the feminine side of us that wants to make our buildings, homes, furniture, and landscapes more beautiful and varied, and wants to bring music, art, celebration, entertainment and joy into our lives. When the masculine side has dominated societies, the innovations have been oriented to needs, and have discouraged maternal innovations, resulting in grim, humorless governments. Fascistic and communistic societies come to mind here, with their military uniforms and Mao jackets, unimaginative buildings, and formalized military displays. When the mother of invention is allowed to provide things that people want, the societies become vibrant, people constantly expand the scope of what they want, and the economies thrive. Today product innovation is driven much more by wants than needs. We want smartphones, computer tablets, large TV screens, and good sound systems. We want to travel and spend our weekends and vacations having fun. While our basic needs could be met simply and inexpensively, we spend much of our income to make our food, automobiles, houses, and furniture fancy, convenient, and attractive. All these require continuous improvements and innovations. Mother drives our successful economies, not father.

We know that inventions and innovations are not always good for humanity. Innovative skills can be used for evil and antisocial purposes, and the development of highly destructive weapons has raised serious moral issues. Another example, the possibility of cloning human beings, also raises moral issues: just because we can develop something doesn't mean we should. Having said this, however, innovation is, on balance, a positive endeavor; it is the most important engine there is for economic progress, and it is essential for improving our lives.

Chapters 1–6 in this book provide a brief history of innovative activity and describe the political and geographic environments. Chapters 7–14 analyze the societal conditions that encouraged innovations, and these are synthesized into a compact set of the essential conditions that support innovative activity. The final chapters provide a look at some great innovators and explore how different kinds of societies do or do not provide these essential conditions.

Part I: Innovation in History

2. Innovation in Early Societies

Sticks, shovels, and streams; Oog helps out the ladies, or not; blue-collar and white-collar innovation

Even though their names are not known, there were innovative people in ancient times whose inventions changed the course of history. Some individual noticed that clay objects painted with colored dyes after firing lost their color fairly soon and thought some dyes might be cooked into the clay and perhaps last longer. Once that idea was formed, people tried out different dyes and learned how they resulted in different colors after firing. Something motivated people to try out other materials, and in trying to imbed colorful minerals in the clay, they found unexpected colors emerging after firing, and the concept of glazing was born.

Similarly, the concept of irrigation came about through a series of small steps, something like those described in the following imaginary story:

> Some individuals watched rivers and noticed that water always flowed downhill. They also noticed after a heavy rain that soil was washed away, temporarily forming smaller streams. They noticed that dirt, branches, and fallen trees slowed the flow of water in streams and caused the formation of a small pond above the obstacles, leading to the idea of damming a stream or river. These concepts languished as children's games for many years until some enterprising individual, concerned about the effects of a drought condition, figured out that by forming a dam of trees at a certain point in the stream the concept could be put to use. In some places the water rose above the dam and filled low pool-like areas with water, and these areas remained filled even after the water flow had diminished. A farmer used this pool to provide water for him and his animals to drink. An adviser to the king, upon seeing this in action, saw that this could take place on a larger scale by digging pools connected by ditches to a small river. These could be dug by using slaves and damming the river, thus collecting enough water for the king's use during the dry season. But how would one build something in a river that would have the same effect as a tree falling across a stream, and how would it stay in place and not be carried by the river's current? It must involve heavy objects like large rocks. So slaves were brought in to drag hundreds of large rocks and to push them into the river from both sides, and the pile of rocks eventually spanned the river. During the dry season, the slaves used smaller rocks and pieces of wood to fill the gaps. During the next

> wet season, water rose above the dam, the pools were filled, and the adviser to the king was set for life. Later, some observant individuals realized that the ditches connecting the newly formed pond to the pools could be expanded into a network of pools that could be located quite a distance below the pond, to where slaves could carry pots to water crops, so food could grow more abundantly. Once having started on such a project, people experimented with different sizes and shapes of ditch networks and found ways of lining the ditches so they would last longer and hold water better. The concept of irrigation was born.

Innovations of the ancient world came about when observant individuals made use of existing knowledge, and improved or expanded on that knowledge to create something beautiful or to meet a recognized need. Larger projects like irrigation required considerable manpower, usually slaves, and so were sponsored by kings or other rich men. As a consequence, the larger projects required a sizeable population with some form of central government.

Stone weapons were developed early in man's history, and evidence of artistic activity in caves dates back several millennia. It's clear that since the earliest times men and women have used their extra time to tinker, explore, and experiment with the materials around them. As Conner points out,[1] it was almost certainly women who developed the knowledge of plant characteristics that were essential for gathering and preparing food. Since they were typically the gatherers, women collected many medicinal herbs, roots, and leaves. The existence of both medicine men and medicine women in tribal societies today suggests that both sexes participated in finding cures for common ailments. They developed an oral encyclopedia of useful plants, and this knowledge was carefully memorized for tribal survival. Songs, chants, and poetry were used to accurately pass this knowledge from one generation to the next.

It is also likely that women discovered that clay becomes hardened after heating in the fire because they typically took care of food-related activities while men were out hunting, fighting, and scouting new areas. For the same reasons, women most likely developed ways of molding clay into forms that would contain food and liquids and eventually invented the pottery wheel.[2] They probably invented jewels and found ways to paint clay forms to make them more attractive. Otherwise, conversations like the following would have taken place:

> Women: Gee, Oog, instead of going off hunting today, could you stay home and figure out a way of shaping clay so that we can store water and protect it from bugs?
>
> Oog: Sure, ladies, we've got enough food for a week, so I can help out. Let's define the system requirements and come up with a working solution.

Not likely. While the descriptions of superwomen innovators in some popular books are probably not accurate, the opposite assertion—namely, that only men were innovators—is certainly wrong.

Since there were thousands of small hunter-gatherer communities, some quite distant from and inaccessible to each other, many early innovations were developed independently in many different places. The methods were communicated from father to son, mother to daughter, elder to younger. For example, the concept of irrigation was independently developed in many different locations; in each location, it evolved as a sequence of small steps, each one unremarkable, but impressive in its final embodiment.

An important lesson to be drawn from this is that some kinds of innovative knowledge were handed down from generation to generation by illiterate people. Specifically, much medical, chemical, and metallurgical knowledge was communicated in this manner until a few hundred years ago.

Once civilized societies began to develop, specifically in the Middle East about ten thousand years ago, other kinds of innovation occurred, involving more abstract concepts like calendars, numbering systems, bookkeeping, and writing. With these came specialists who made their livings by mastering these skills. They became astronomers, scribes, and priests. They served as advisers to kings and emperors and as religious leaders of the people. With the development of writing, these concepts were often handed down by writings on clay tablets, papyrus, and scrolls. These specialties were often jealously guarded by the family groups involved in order to maintain their influence over time.

Another phenomenon was echoed in many other aspects of civilized life: a class structure of haves and have-nots, nobility and peasantry, educated and uneducated, Brahmin and untouchables, privileged and underprivileged, aristocrats and slaves, lords and commoners, literate and illiterate. The privileged came to consider themselves above manual endeavors. In intellectual areas, this was reflected in the separation of theoretical and practical. Pure thought became considered superior to grubby hands-on knowledge, a condition that was to have profound consequences for the development of modern science.

This phenomenon also had important implications for our subject because it resulted in two different sources of innovation: that of practical hands-on innovation and the sort that involved abstract concepts and principles. For the sake of highlighting the differences between these two, I refer to them as blue-collar and white-collar innovations. Most of the innovative activity prior to the development of civilized societies would be considered blue-collar, where the knowledge was handed down primarily by oral means. White-collar knowledge increasingly came to be handed down through the written word. Reading history books today, one would get the impression that once civilization was developed, white-collar innovation essentially supplanted blue-collar innovation. Many authors who have studied the history of technology would strenuously object to this, and their books show many counterexamples of blue-collar innovations throughout history. *It would be more accurate to say that most inventive activity was of the blue-collar variety up until the 1700s and developed quite independently of any white-collar activity.* At about that time, the two kinds of activity began to become mutually dependent. This dependence became stronger, and by now the distinction between them is less meaningful. Today the terms "science" and "technology" are better descriptors. This will be discussed in more detail in chapter 6.

There is conclusive evidence of white-collar innovative activity occurring as far back as 5000 BC, and, of course, the onset of agriculture and the domestication of animals date back several thousand years before that. Some of the best records of today for that ancient period come from Egypt and Mesopotamia, where, again, both blue-collar and white-collar innovations took place.

Egypt[3]

Predynastic Egyptians between 5000 BC and 3200 BC fashioned tools, weapons, and ornaments from flint, copper, silver, and gold. They developed new processes for finishing, glazing, and decorating pottery, and they developed efficient irrigation systems and reclaimed swamplands. Methods were formulated for weaving linen of high quality. Predynastic Egyptians invented the first solar calendar, one that had twelve months of 30 days each, with 5 days added at the end of the year, for a total of 365 days. They invented hieroglyphic writing and began building the pyramids, which certainly required significant innovations in stonecutting, transportation, and building techniques.[4]

In the age of the pharaohs, from 3200 to about 1150 BC, Egyptians developed the concepts of addition, subtraction, and division. Multiplication was performed through an interesting series of additions.[5] They developed a form of the decimal system, but without a zero. Mathematicians computed areas of triangles, rectangles, and hexagons and volumes of pyramids and cylinders. They calculated the ratio of the circumference of a circle to its diameter as 3.16. We now know the exact ratio is π, approximately 3.14159, which means the Egyptian mathematicians were accurate to about 1 part in 150—pretty impressive. Astronomers mapped the locations of the stars. Egyptians invented the sundial, papyrus, glass, and some metallurgical processes. In about 3000 BC, they invented a kind of alphabet, although it was the Phoenicians who much later developed an independent alphabet exclusively for communicating.

During this period, Egyptians built irrigation canals[6] and dams. They built ships to sail the Red Sea and the Mediterranean, manufactured pottery, glass, and textiles, and developed techniques to cut and shape hard materials before iron tools were developed. In an exhibit featuring Hatshetsup, the first female pharaoh, I counted twenty-one different materials used for statues, buildings, and decorative objects.[7]. In building the pyramids, the Egyptians developed techniques for cutting stones, moving them into place, and shaping and aligning them to surprisingly precise tolerances. They invented musical instruments: the harp, the lute, and woodwind instruments similar to oboes. To support commerce, they developed written orders, receipts, property deeds, contracts, and wills. They had no coins, but rings of copper and gold served as a medium of exchange. There were medical innovations: Egyptians developed dentistry and surgery and compiled a catalog of medicines. They had some understanding of the heart and stomach and treated fractures and performed simple operations.[8]

So over a period of four thousand years, the Egyptian civilization introduced an impressive number of innovations, many of which have endured to this day with surprisingly little modification. Of course, what is also impressive is Egypt's sheer longevity as an identifiable, coherent civilization. Four millennia constitute a span of time difficult to comprehend today. Spread over such a long time, the rate of innovation is perhaps less impressive, but changes did occur throughout the entire period. Cumulatively, the number of new ideas, concepts, processes, inventions, and arts is quite awe-inspiring.

Mesopotamia[9]

The Sumerians, who lived in Mesopotamia from about 4000 BC to 1800 BC, may have been the first people to invent a writing system. Theirs was known as cuneiform and consisted of wedge-shaped characters imprinted on clay tablets. Originally, the symbols were not based on the sounds of the language, but the writing system evolved to include symbols for some syllables and grammatical elements. This was quite an impressive feat. As Jared Diamond points out,[10] inventing a writing system from scratch was much more difficult than borrowing from or adapting a writing system already in existence. Cuneiform texts were not limited to account ledgers and property records, but included poetry, lullabies, astronomical events, and lists and descriptions of animals, medicinal plants, and stones.

The Sumerians developed processes of multiplication and division and of computing square and cube roots. They and the Babylonians that followed developed a place-value system like our current one, where a digit on the left has a value of ten times the value of the same digit to the right. They used a base sixty, however, rather than ten. (We also sometimes use base sixty, in a sense, with our sixty minutes per hour and sixty seconds per minute for measuring both time and angles.) The Sumerians also invented the water clock and the lunar calendar of 360 days, based on cycles of the moon.

Sumerians invented the wheel in about 3500 BC, which they used in pottery wheels and carts. They worked wool into cloth and flax into linen. They built canals and dikes to control river water. By about 2500 BC, they had developed glazed surfaces and freestanding glass objects. They did metalwork, gem carving, and sculpture. In their metalwork, they first learned to purify copper through heating and later worked with bronze and then gold and silver. They imported their metals, woods, and lapis lazuli; they exported textiles, jewelry, and weapons.[11] Sun-dried brick was used for building the ziggurats, since appropriate building stones were scarce. For their royal tombs and private houses, they invented the dome, the arch, and the vault.[12]

The Assyrians existed as far back as 3000 BC and began to expand their territory in about 1300 BC. They took over Babylon in the tenth century, and their empire reached its peak from 722 to 612 BC. Their interests were largely military, but they did have some scientific and medical accomplishments. They divided the circle into 360 degrees and had a concept of longitude and latitude. They attempted to predict eclipses, with some success. They catalogued more than five hundred drugs. An enlightened leader, Assurbanipal, established a royal library that contained more than twenty-two thousand tablets.[13] The Assyrians had sophisticated knowledge of minerals and rocks and understood that copper had both black and red oxides. They knew about acids. They acquired ironworking from the Hittites and were the first to outfit armies with iron weapons. They invented movable towers, battering rams, and other siege engines and made good use of chariot designs.[14]

The Chaldeans defeated the Assyrians in 612 BC and established a new culture that lasted about a century. During their tenure, they made some advances in astronomy, where they developed detailed maps of the stars.[15] A most impressive engineering feat was that of the Hanging Gardens of Babylon, which King Nebuchadnezzar had designed and built for one of his homesick foreign wives. His clever architects used suspended terraces, and the gardens were watered with screw-like devices that lifted water from the Euphrates.

The Mesopotamian innovations were developed primarily to provide the ruling authorities with capabilities they needed, namely military, medical, astrological, and agricultural applications. Although the innovators may have been well-known to their contemporaries, their names didn't make it into the written records. That would come later.

Asia

Two other centers of civilization existed during this same period: one along the Indus valley in India/Pakistan and one along the Yellow River in China. These civilizations developed in size and sophistication, with populations that exceeded those of the West. They, too, developed written languages, calendars, and mathematics. The Chinese developed a decimal place-value system and invented the abacus based on it.[16] In astronomy, they made careful observations and measurements, estimating the year to be 365.25 days long (compared to the currently accepted value of 365.24). They developed maps of the stars and techniques to predict celestial events. They developed irrigation canals and built large structures. Their medical knowledge was on a par with that of the Western civilizations.

Teresi points out that the third millennium BC in India was called the "Age of Cleanliness."[17] Some of its cities had toilets and sewers, with lavatories built into the outer walls of houses. The toilets had wooden seats with vertical chutes that dropped waste to municipal drains and cesspools. Even today much of the world lives in less sanitary conditions. The caste system was introduced later, and the Brahmin, or scholar, caste dominated the development of mathematics and science.

With individual exceptions, the inventions and developments that occurred in Asia were very similar to those to the west, reflecting the values and needs of "hydraulic" civilizations. They also reflected the values of highly centralized governments.

Governments of the Time

The governments of the larger societies during this time were mostly autocracies, headed by emperors or kings who had virtually unlimited power over the people. Each of them was supported by a group of powerful families, who vied for his favors and ran the economy. These families managed the workers; headed up the military and fought the wars; designed the buildings, cities, and tombs; and managed the multitude of slaves, farmers, and craftsmen who did the everyday work. These families also provided the priests and counselors to the autocrats. It was common for the autocrats to control the state religion and to claim that they were gods themselves or at least chosen and blessed by the gods. The economies were dominated by state-run monopolies. The kingdoms traded with each other and fought over resources. The rulers vied with each other to build the most impressive cities and structures and demonstrate their power. Inventions were put to use if they supported the autocrats' concerns for controlling the populace, supporting military adventures, and constructing and decorating buildings and tombs. While the rulers had a stake in making sure the populace had enough food to keep from starving, they had little stake in spending royal funds to enhance their welfare beyond a subsistence level. As a consequence, there was little innovative effort to improve the lot of the lower classes.

3. A Different Kind of Society Appears on the Scene

The Greeks develop the revolutionary concept of citizen participation in government; the two greatest innovations of the Greeks; innovation around the Mediterranean; the Romans lose the spirit

While the ancient kingdoms and empires had similar autocratic forms of government, they varied widely in their size, resources, and lifetimes. Empires rose and fell; kingdoms came and went. Large strong kingdoms conquered small weak ones, and large weak kingdoms were attacked by small strong ones. Large kingdoms had the advantage of a common language that gave them staying power and the desire to reunite after being torn apart by war. However, small kingdoms had advantages of their own in some places, particularly if they were geographically separated from the large empires. The leaders of the larger kingdoms often found it more convenient to trade with them than to conquer them, particularly if they paid proper homage. City-states developed along the Mediterranean with some independence from Egypt and the Middle East, notably those of the Phoenicians and the Greeks. Partly because they were so small, and partly because of their proximity to the sea, the trade between them was quite active. This resulted in the development of a merchant class that acted to counterbalance the power of the kings and their loyal families; in the more powerful autocracies, the rulers made sure the merchants were their agents. In many Greek city-states, the kings were weak enough that a new society emerged, one that empowered a much larger group of people. These people, called citizens, gained enough power that they established several revolutionary institutions and a new political idea: democracy.

The Greeks developed two innovations that I claim have not been given their proper recognition as groundbreaking concepts: (1) the stadium, where citizens could watch, cheer, and gamble on athletes competing in races and other contests, and (2) the amphitheater, where playwrights and actors could entertain the citizens with stories that addressed their common interests and that even mocked the aristocracy. They also created the idea of the "individual achiever," who could achieve fame without pandering to the autocrats. These were revolutionary developments. They were not accepted immediately by the rest of the world, and two thousand years would pass before the idea that government derives its power from the consent of the governed would become popular, a concept solidly based on the Greek experience.

The Greek experience lasted only a few centuries, and the Roman experience that followed adopted many of its democratic ideas initially. However, the autocratic urge eventually reemerged, political power became highly concentrated, and the Empire replaced the Republic. Democracy disappeared from the world scene. The Greeks did leave an enduring legacy, however, in their writings and in establishing forever the concept that the interests and desires of the populace matter.

Greek Societies[1]

Because so much has been written about the ancient Greek societies (which included the coast of present-day Turkey and much of Italy and Sicily) and so much praise has been lavished on them, discussing the ancient Greeks with any objectivity is difficult. A number of European historians have erroneously given them credit for the development of many mathematical and astronomical ideas, even though they inherited much of their mathematical and astronomical knowledge from the Egyptians, as they themselves acknowledged. It's also true that the Babylonians knew about the geometric properties of what would eventually be called the Pythagorean theorem hundreds of years before Pythagoras.[2] Furthermore, the Greeks didn't really invent the idea of the alphabet—it was invented by Canaanite and Hebrew Semites and propagated by Phoenician merchants.[3]

But Greek societies, which included cities on the coast of present-day Turkey and much of Italy and Sicily, were unique. For the first time, at least in the accounts we have access to, we learn not just about rulers and their exploits but also about historians (Homer, Herodotus), playwrights (Aeschylus, Sophocles, Aristophanes), and philosophers (Pythagoras, Socrates, Aristotle). Philosophy then included the subjects of science, music, and mathematics. The records of their individual accomplishments, attributed to them by name, set an example that changed the world forever. We hear their ideas coming from their own voices, free from the strictures of the egos of rulers and the orthodoxy of priests. Their voices speak to us directly across the centuries. We can argue with their ideas, and we can gasp with amazement at the depth of their reasoning. Even though the numerous translations and retranslations of their works have no doubt introduced some distortions, a sense of personal connection comes through that is totally lacking in prior writings.

This was helped by the Greek language itself, which introduced vowels into a small alphabet that was thus accessible to a wide range of people and supported an expressive new vocabulary.[4] But more important enabling features of the Greek society were (1) the several freedoms of its citizens—to travel, to pursue different lines of work from their parents, to participate in political life, to be entertained—and (2) the weakness of religious and political limitations on the exploration of ideas and on individual initiatives. While many of the great intellectuals were from noble families, many were from modest families; for example, Socrates's father was a stonecutter, considered a lowly profession by Greek citizens.

The Greeks also enhanced the concept of drama, the portrayal of the human experience by actors on a stage. The intimacy and immediacy of the dramatic experience was vivid: the shape of the theaters made it possible for a large audience to hear even the breathing of an actor on a stage. What is less appreciated is that the actors themselves felt the presence of the audience with an equal intensity. Every reaction from those on the steps—sympathy, laughter, and disapproval—was immediately and strongly sensed by the actors, a daunting interaction. You can feel this when you visit an open-air theater in Greece or Sicily. Drama was copied all over that part of the world and, of course, is still enthusiastically experienced around the world today. Drama enabled the propagation of political and social ideas beyond the control of the ruling classes.

Athens was also blessed to have three enlightened aristocrats in the early days who established unique public institutions over the period of a century: Solon, Cleisthenes, and Pericles. Between them, they established councils and assemblies that involved citizen participation, with powers to declare war and appropriate money. They set up courts that reviewed disputes and passed judgments, also manned by citizens. No longer would it be universally assumed that rulers inherently held these powers. The concepts of executive, legislative, and judicial branches of government were given a democratic meaning. Citizens flocked to Athenian hills such as the Pnyx to debate and cast stone ballots to vote on issues.

While the Greeks are credited with creating the first democracy, not everyone participated in public life. Cahill[5] believes that, taking into account slaves and freemen in and around Athens, only about one out of five people were citizens. He estimates that "at their height, Athenians probably numbered no more than a quarter million, of which as many as 100,000 may have been slaves." Since women did not participate equally in public affairs, the Athenian democracy could not be considered a true democracy by modern standards. Nonetheless, the direct participation of so many people in their own government and in public life was a radical departure from anything else that had ever existed. These citizens also had something else very important: time. They had the luxury of being able to support themselves and still have time for other activities. They had time to participate in the assemblies and courts. They had time to debate political issues. They had time to learn about philosophical and social issues, and they discussed them for hours. Individuals openly took responsibility for their lives.

What was it about the Greek society that made it different from others in that part of the world? Why didn't such an explosion of ideas happen elsewhere? Historians are still arguing these questions, but they generally agree on some relevant factors.

First of all, the invention of a small alphabet was important, because most people could learn it without years of specialized training. Cahill[6] claims that this alphabet and the resulting expressive vocabulary had a democratizing effect and empowered the populace in a way that was not previously possible. Earlier writing systems were so complicated that only a specialized priesthood like the biblical scribes could use them. Second, by virtue of its distance from the large autocracies, Greek society evolved from small monarchies to city-states. These simple words belie the huge significance of the changes. Kings and rulers usually owned the land and the buildings, and, in a sense, they owned the people who lived within the kingdom—the people were subjects, not citizens. The idea that an individual could propose a law or course of public action was unknown—the king (with the help of advisers) made all relevant decisions. But this idea evolved in Greece. It didn't happen overnight, of course. Greece had kings at first, but they did not have the absolute power of a Nebuchadnezzar. Councils of nobles and assemblies of warriors were created to advise the kings, but they also acted as a check on the kings' power. Kings were replaced over time by dictators (called "tyrants," who were actually elected by citizens), but these rulers never achieved the absolute powers of the Eastern kings, and eventually democratic institutions developed. Third, the aristocrats didn't own all of the land and never developed levels of wealth that made them vastly more powerful than the average farmer. (Early Greece was largely an agrarian society). One reason was that trade developed and grew between cities in the Mediterranean, resulting in a large number of prosperous merchants. These merchants had interests different from those of the landowners, and the resulting division of power served to check the power of either group. For these reasons, the Greek society escaped the despotism of kings, developed a middle class, and became much more egalitarian and open than other societies of the world.

I claim that there are three other concepts that emerged from the Greek experience, ones that are rarely cited in the historical accounts with more than a passing reference, concepts that arguably have had more influence on our world than all the intellectual products of the great philosophers, poets, and writers. These concepts are (1) the individual achiever, (2) the athlete as hero, and (3) competition, ideas that developed concurrently and played off of each other. Each of these was significant for innovative activity.

The concept of the individual achiever, the notion that individuals can be and should be given credit by name for their ideas and accomplishments, was born in Greece. With this concept came the notion that people are not merely subjects of the ruler but capable of inventing things on their own and coming up with ideas on what is important, outside the narrow concerns of the autocracies. Moreover, they deserve to be recognized and esteemed by their fellow men, like the historians, playwrights, and poets mentioned earlier. We know little and do not care about the status of their families in the society, how many troops they commanded, or how magnificent their houses or tombs were. Their importance is not tied to the accomplishments of their ruler. This represents a momentous change from the other societies of the world at the time. The Greeks showed that a wide variety of people, not just the aristocracy or the priesthood, could develop new ideas and become politically active, and thus make meaningful changes in their own societies. This is the concept of the individual achiever, an idea that would never be entirely suppressed.

Second, in the Greek society individual athletes could distinguish themselves by extraordinary feats and achieve fame and recognition in the eyes of their fellow citizens. By running the fastest race, a young man could become famous among his peers and be recognized in the streets of the town and praised for his skills. Competitions first involved running and were later expanded to include jumping, lifting, wrestling, and throwing (heavy stone weights, for example). People also competed in races using animals. Contests involved both individuals and groups. Once the desire to compete caught on, the Greeks were forever inventing new contests. These events became enormously popular, and soon stadiums and coliseums were built to stage the events. Prizes were awarded, and people applauded their favorites and gambled on the participants. Crowds flocked to the stadiums, and the events became an important part of people's lives. The concept of athletic events, where ordinary people come together to shout, cheer, and have fun, was born, and public leisure time activities were no longer solely the domain of the aristocracy. This, too, was a profound change that had enduring consequences. Today athletic events are held all over the world, with few fundamental changes from the original Greek concept. From the standpoint of the sheer number of people affected and the amount of time and energy involved, the influence of athletic events on people's lives today is much greater than the musings of the philosophers. Once the concept was developed, it enjoyed widespread and lasting acceptance. A ruler today who dares to eliminate such activities will not last long. Furthermore the concept is subversive to autocracies—the people are entitled to leisure activities, and they can't be expected to spend all their lives working for the benefit of the ruler.

Third, the concepts of athletic events and recognition of individual efforts go hand in hand with the concept of competition—not the limited form of competing for the favors of the autocrat, but competing in an area where the participants individually gain something important to them: status, recognition, fame, money. Competition was pervasive in all their endeavors. According to Cahill,[7] the athletic competitions that spawned the Olympics were just part of the story. Playwrights and poets competed for prizes at the amphitheaters, athletes competed at the coliseums, and scholars competed to teach young people. They were widely known and admired by the populace. This recognition afforded them power that would have been unthinkable in Mesopotamia—they could criticize the aristocrats and get away with it. This power was crucial to the birth of democracy.

While the works of historians, playwrights, poets, and scholars are white-collar efforts, the Greek society developed creative blue-collar innovators as well. No doubt the athletes found ways of making themselves stronger by inventing exercises and healthful diets. The lowly artisans were looked down upon, even despised, by the citizenry. Even they, however, competed fiercely with each other to produce the most beautiful pottery, as attested by inscriptions on vases, pitchers, and other pieces recovered from archaeological digs. By the way, it is interesting to note that these pieces we value so highly were made by people who were considered by the citizens of the time to be the riffraff of society.

This idea of the individual achiever did not immediately spread around the world. It was confined to a few cities in Greece and some of its colonies. It did not exist at all in nearby Sparta, and it did not guarantee Athens' survival; Athens fell to the Spartans in the Peloponnesian War

(431–404 BC). Later the Romans took over Greece. The Romans, while they admired the Greeks and retained some democratic political institutions during their republican phase, focused on military, administrative, and political affairs and de-emphasized philosophical pursuits. The Greek ideas were not, and are not, popular with powerful people. Indeed, in plays by Aristophanes and Euripides, the aristocrats of Athens were made fun of, much to the delight of the audience. Such lack of respect was not likely to prove popular to neighboring kings. In fact, rulers have always resisted the democratic ideas embodied in some Greek writings. Control is never given up without a struggle.

It's worthwhile to look more closely at Sparta[8] because of its dramatic contrast with Athens and its colonies and with the other city-states. Sparta had about the same population as Athens, and its people were racially the same—they came from the same Dorian stock as the rest of Greece. It, too, had governmental bodies that distributed political power, at least on paper. But there were significant differences. Sparta had two hereditary kings who had equal power, which meant in practice that they had difficulty acting effectively together. Real power was in the hands of a group of five men called "ephors," who exercised virtually total control of the populace. Although they were elected by a popular assembly of citizens, once they were in office they were untouchable. They only served for one year, however, which kept any one family from gaining too much control. Discipline was maintained by a tight set of unwritten agreements and an all-pervasive religion. The Spartan society had very strict roles for everyone. The upper class was strictly military, and military considerations trumped everything else. The kings acted as generals, which kept them occupied, while the ephors controlled the laws and the religion. Everyone's lives were prescribed. Each boy citizen went into the military at the age of seven, not to leave until he was sixty years old. At age thirty, he was married off and allowed to see his wife long enough to get her pregnant, and that was about it. Meanwhile, for each citizen there were about seven serfs (called "helots") to do the menial work.[9]

Other than the techniques they used in their military prowess, the Spartan population produced virtually nothing in the way of ideas or inventions. The Spartans had no concept of participatory politics. Unlike the Athenians, they were not stimulated by ideas that came from trade and interactions with other people. This intellectual isolation was reinforced by, and perhaps resulted from, their geographical isolation from other societies—they were surrounded by hills and had no good harbors to support a trading fleet. There were no rewards for inventions for other than military applications, and the highly constrained lifestyle made sure there was little competition of ideas. Improving the lot of the populace, even of the citizenry, was not a societal value. But they were formidable fighters and deft strategists and tacticians.

By the time Aristotle was born in 384 BC, Greece had changed significantly from the glory days of Socrates. By then the Spartans had been defeated by the Thebans, who in turn were defeated by an alliance of Greek city-states. So while Athens retained much of its cultural heritage, it was no longer the democracy and the center of influence that it had been before. However, its ability to produce men of genius was not yet spent. Aristotle continued the tradition of the individual teacher/writer/thinker, one not beholden to government or political interests. Born into wealth in northern Greece, he went to Athens as a young man to study with Plato, who was born during the Peloponnesian War. He became a member of the Academy and founded the Lyceum.[10]

Influenced by the attitudes of the theoretical scholars of the times, Aristotle deliberately concentrated his research efforts on subjects that had no obvious practical value: biology and natural philosophy. He was especially interested in developing cosmological explanations that appeared to fit with what his senses observed. His writings would later dominate much of European thinking, but it's important to recognize that his contemporaries were not nearly as awed of him. Theophrastus, his student, and Strato, who succeeded Theophrastus at the Lyceum, wrote quite critically of some of his concepts.[11] Both men probably would have laughed at the quasi-deity status attributed to Aristotle by the Catholic Church during the Middle Ages.

Greek white-collar innovation in the period of Plato and Aristotle was concept oriented, not observation oriented. The Socratic method of inquiry opened up new debating possibilities. Abstractions such as ideas and forms were given an importance they had not had before. The Greeks developed the concept of abstract mathematics—the notion that certain truths exist in the abstract, independent of physical manifestations, which can be applied to the real world. Even an all-powerful god cannot change the fact that two plus two equals four. The truth of the assertion is inherent in the definition of the numbers.

The effect on Greek scientific thought was not totally positive. The stunning power of reasoned thinking to explain the world and human behavior had the result of de-emphasizing observations. The sneering attitude of Plato and other scholars of the time toward experimentation had the effect that practical applications and careful observations were not considered particularly worthwhile. Intellectual thought was valued to the point that the philosophers actually discouraged experimentation. Aristotle did swim against this current somewhat, particularly in his studies of biology, as he spent several years studying animals and birds. But this effort was an exception. The result was to hold back developments in astronomy and physics for almost two thousand years. Later, scholars reading the great Greek philosophers were so impressed by the depth of their reasoned thinking that they accepted their conclusions, which we know now to be totally fallacious: the earth is not the center of the universe; the sun, planets, and stars do not orbit the earth mounted on perfectly round celestial spheres; everyday objects do not send out rays of light to make themselves visible to us; and motion does not require a continually applied force. To state what is now obvious, deeply reasoned thinking based on faulty assumptions results in erroneous conclusions. The world had to wait for Galileo in the early seventeenth century to start turning this around. In summary, white-collar scientific innovation suffered a severe setback in Europe.

Except for Aristotle, the Hellenistic innovators that succeeded him came less from Greece than from the cities and islands around the Mediterranean that had come under the influence of Athens: Pergamum in present-day Turkey, the island of Samos off its coast, Memphis in Egypt, and Syracuse in Sicily. The center of innovative activity moved to the remarkable Alexandria, which fortunately did not fully subscribe to the Athenian aversion to the physical world.

Alexandria was built immediately after Alexander's conquest and was named after him. He set up a Greek as king of Egypt, Ptolemaios Soter, a man who supported science and learning.[12] His successor, Ptolemaios Philadelphus, founded the Museum at Alexandria around 280 BC, an institution that would last for seven hundred years. Shortly afterward the famous Library of Alexandria was added, and it would soon contain five hundred thousand scrolls. This was the first institute for advanced research, a place where scientific and literary scholars from all over the Mediterranean gathered to conduct what we would call "pure" and "applied" research. They received money and food from their royal sponsors and were allowed unusual latitude in pursuing their interests (which did not make them particularly popular with the populace).

One of the most famous of the museum's early members was Aristarchus of Samos,[13] who concluded that the stars looked like they remained in one place because of their immense distance from us and deduced that the earth and planets revolved around the sun. As this seemed counter to everyday observations and the teachings of Aristotle, his model was not accepted by his colleagues, and it took almost two thousand years for his conclusions to be widely adopted. Since this also flew in the face of existing religious thought, he was threatened by local religious authorities.

Another famous astronomer at Alexandria was Apollonius of Perga, who developed the tools of epicycles and eccentrics to explain planetary motion. Several hundred years later, Ptolemy, also at Alexandria but when it was under later Roman rule, would base his famous explanations of planetary motion on these ideas and develop a more complete model.

Hipparchus of Alexandria invented the astrolabe, or at least developed a working instrument, and came up with a reasonable estimate of the diameter of the moon and its distance from the earth.[14] Eratosthenes of Cyrene estimated the circumference of the earth to better than 20 percent accuracy by observing the difference in the angle of the sun at noon at the summer solstice between Aswan and Alexandria and estimating the distance between the two cities. He was one of the first to claim that it might be possible to reach India by sailing west, as he believed that all the oceans were connected.

At Alexandria, Euclid wrote his definitive *Elements of Geometry*, a book that was considered the classic text on mathematics into the nineteenth century. While most of it was a compendium of the work of other mathematicians, it was written in a concise formal and abstract style that would set the tone for later scientific writings.

Archimedes, the great Hellenistic physicist and engineer, lived and died at Syracuse but visited Alexandria and communicated with Eratosthenes. He developed the concept of specific gravity, as well as the principles of the lever, the pulley, and the screw.[15] Later, Hero of Alexandria would be credited with a wide range of inventions, including a predecessor of the church organ, a steam engine, an alignment device for surveying, and a siphon. He was a gadgeteer who delighted in designing special theatrical effects.

Significant developments also took place in medicine at Alexandria, continuing the tradition of the school at Cos that had thrived centuries before, a tradition that included Hippocrates. Herophilus of Alexandria was one of the first to practice human dissection, and he discovered that arteries contained only blood, contrary to Aristotle's claims that they contained both blood and air. Erasistratus discovered the valves of the heart, and he claimed that arteries and veins are connected. He rejected the humoral theory of disease and taught against bloodletting; unfortunately, while he was eventually shown to be correct, humors were considered real by medical practitioners for two thousand years, and bloodletting is still performed in some areas of the world today.

Alexandria was not the only Hellenistic center of learning. The Academy and the Lyceum at Athens continued to operate, and Pergamum had an impressive library and state-supported scholarship. But it was at Alexandria where a number of great minds worked and interacted with each other, challenging the conclusions of the great minds that came before. The shame is that so much of the observation-based knowledge that was developed at Alexandria was unknown to (or ignored by) later scholars, or was destroyed.

The Roman Empire

For two hundred years after the Peloponnesian wars, the Mediterranean area was fragmented. This ended with the emergence of the Roman Empire. By 200 BC, Rome had consolidated and was sending out armies to establish an empire. This empire was to last about six hundred years.

It is puzzling that in spite of being heir to the Greek intellectual heritage and continuing the tradition of a citizenry, the Roman society was not highly innovative in the white-collar sense. The Romans developed a short alphabet similar to that of the Greeks, and they admired Greek accomplishments and built Greek-like coliseums and theaters. They adopted the Greek gods. They continued the Greek tradition of supporting drama, although at Taormina in Sicily the small central stage was removed and replaced by a larger area so that gladiator-type events could be held. Coliseums were built to entertain the people with fights between people and animals. But they made no improvements in mathematics, astronomy, or what we call natural sciences. For example, Pliny the Elder wrote a book about the natural world but didn't distinguish between sound facts and hearsay, giving credence to stories such as those of humans with feet that turned backward.[16] According to Teresi, "Ancient Romans had not kept alive the scientific inquiry of the Greeks, and medieval Europe, out of touch with the classics, had little scientific tradition of its own."[17]

One plausible explanation as to why the Romans didn't continue the innovative tradition of the Hellenists is that few relevant works were translated into Latin, so that most young Romans were not aware of the ideas that had been developed.[18] This argument is compelling and goes a long way to explaining the lack of scientific innovation by the Romans.

Apparently, scholars primarily translated plays, religious books, and political writings, but not mathematics and science texts. The Romans seemed indifferent rather than hostile because they didn't interfere with the schools at Alexandria, Pergamum, and Athens, and they even supported them to some extent.

Another possible reason why these writings weren't considered important is that they probably weren't considered important even by most Greeks and Hellenists, who likely viewed the scholars as an oddball group who constantly argued among themselves over matters of no great importance to everyday life. It wasn't until several centuries later that the writings of the Greek philosophers were perceived as the works of geniuses. When Socrates publicly chided the Athenians for their conduct during the Peloponnesian War, he was killed for his opposition. No influential group in Athens came to his defense, which suggests that the scholars did not have a lot of public support.

Medical advances took place within the Roman Empire, but they were accomplished by individuals trained in the Hellenistic schools. The most famous was Galen of Pergamum, who served as court physician to the emperor Marcus Aurelius in Rome. He wrote prolifically about anatomy and physiology, developing models of the human body that served as the definitive reference for about 1500 years.[19] Unfortunately, he also reinstituted the humoral theory and bloodletting that Erasistratus had rejected.

The Romans did show creativity in other ways. They experimented endlessly with bureaucratic and political offices and established new legal and governmental concepts. They established representative bodies (the Senate and the plebeian assembly) and magistrates (consuls, praetors, aediles, and quaestors). They established an elaborate system of checks and balances, although these were not enough to prevent the despots of the late empires. Their generals showed great abilities to adjust tactics to defeat their enemies. They were skillful in running a far-flung empire and at co-opting local leaders by making them Roman citizens.

They put their technical skills into engineering projects, where blue-collar innovations took place. One useful invention was that of the screw press, which they used for pressing grapes and olive oil. More important, they invented concrete and the process for pouring and curing it, and they put this material to good use because limestone and sand (the principal ingredients of concrete) were plentiful. They built aqueducts, bridges, roads, coliseums, amphitheaters, and other buildings throughout the empire. Roman roads remained the best available in much of Europe well into the Middle Ages.

So their creative energies went into practical matters. The heroes of a young Roman man were military and political. Writers were most respected if they were political figures like Cicero. The accumulation of wealth, power, and status was the main objective of ambitious people.

Italy had few natural resources other than farming, so there wasn't a lot to build industries with, even if the impulse had been there. Resources of other countries were brought back to Rome, which depended on the empire's products, in the tried-and-true method of plunder.

The Roman emperors spent much of the first four centuries trying to maintain the empire, which everywhere experienced attack after attack. Eventually it succumbed. After Alaric I, the king of the Visigoths, sacked Rome in 410, its subsequent decline was continual, and by 476 the fall of the Roman Empire was complete. Rome itself suffered a decline in population from one to two million people at its peak to less than fifty thousand after its fall. Europe disintegrated into a number of warlords, feudal estates, and villages; the period commonly called the Dark Ages set in.[20]

4. The Mediterranean Loses Its Innovative Spirit, then Regains It

The breakup of the Roman Empire—a blessing in disguise; Asian inventions begin a chain of world-changing innovations; an enlightened Islamic Empire emerges; blue-collar innovation revives and thrives in Europe; increased trade stimulates innovation

With the glory days behind them, the Mediterranean centers of intellectual activity and inventions declined, and Rome was preoccupied with maintaining its empire. After the fall of the empire, Europe was fragmented into small areas with unstable and changing governments. Centers of Christianity sprang up around the Mediterranean Sea, abetted by the roads and water routes of the Roman Empire, but the Catholic Church as a political force did not emerge until centuries later. After the Germanic invaders sacked Rome in 476, towns and cities were reduced to villages, and there were no effective, stable governments of any significant size. With the Roman citizens dispersed from the city, and with the large number of languages in the area, literacy was drastically reduced. The few literate people with time to ponder were a few scattered clergy, like Augustine, who concerned themselves with religious and philosophical matters. Before the final fall, the Roman government issued decrees that attached peasants to the land they worked, so they were effectively owned by the landowners. This set the stage for the feudal manors that would be the basis of the European economy and political landscape for centuries. This period was characterized by frequent raids by internal and external marauders, competing armies, warrior chiefs vying for control, and small population units. There were few towns of any size, but lots of small villages. Mountains and water separated people from each other. People became isolated, Roman roads were destroyed or left to decay, and communication between people inland became limited. Most trade was by sea, so that communication between different peoples took place primarily along coastal areas. Most people, including the feudal lords, were illiterate. Superstition was rampant. This was not a set of conditions favorable to innovation, particularly of the white-collar variety.

However, in this seeming intellectual desert, the seeds of recovery were sown. When Constantine moved the capital of the Roman Empire to what subsequently would be named Constantinople in about 330, many books were moved as well, so that some ancient knowledge was protected from the later destruction of Rome. This knowledge would become available centuries later to Islamic and European scholars. The growing Catholic Church organization soon began to permit small monasteries, where the monks ran farms to support their religious practices. These maintained and trained a literate clergy, and the small farms grew into large monastic estates that would eventually nourish a flowering of practical innovations. Meanwhile, the very weakness of any central control would enable developments unhindered by the heavy hand of autocracies, with their monopolistic practices and other limits on innovation. Landes noted that this fragmentation "was the strongest brake on willful, oppressive behavior" and that "by decentralizing authority, it made Europe safe from single-stroke conquest."[1] He concluded that "Europe's great good fortune lay in the fall of Rome and the weakness and division that ensued." *Far from being a tragedy for the world, the breakup of the Roman Empire enabled innovation to revive in Europe.* But it would take several centuries before these conditions would bear fruit across Europe.

After the fall of Rome, the new Byzantine Empire was run out of Constantinople, and it controlled much of the Mediterranean that had been part of the Roman Empire. It retained many of the features of the older autocracies—the emperor was "an absolute sovereign with unlimited power over every department of national life"[2] and was considered a representative of God on a level with the Apostles. The government controlled all activities; however, it respected the works of the Greeks and Hellenists and retained many of their writings in libraries. While few technological innovations took place, the Byzantines did build schools and libraries—libraries were centers of learning, and the schools provided instruction in mathematics and medicine. The hospital, a public center of medical technology and treatment of patients, was a Byzantine creation and a significant contribution to the practice of medicine.[3]

Innovation Blossoms in Asia[4,5]

While Europe was in its fragmented state, it had no cities or societal centers large enough to support white-collar innovations like those that had been developed in the earlier autocracies. Intellectual activity was primarily religious in nature. The constant marauding and feuding acted as a brake on blue-collar innovation, and much of the knowledge of the road and building construction techniques of the Romans was lost.

The rest of the world was not idle, however. During the period from AD 300 to 650, one active place for white-collar activity was India under the Guptas. Some of the work performed there prepared the way for later innovations elsewhere. Unlike astronomical efforts in the rest of the world, India actually had six different competing schools of astronomy and astrology.[6] According to McLellan and Dorn, "From the fourth through the seventh centuries various Indian astronomers produced a series of high-level textbooks covering the basics of astronomy: the solar year, equinoxes, solstices, lunar periods, the Metonic cycle, eclipses, planetary movements (using Greek planetary theory), seasonal star charts, and the precession of the equinoxes. Aryabhata I (b. 476 A.D.) lived in Pataliputra, composed a *siddhãnta* [meaning 'solutions'], trained students, and held the unorthodox view that the earth rotates daily on its axis." He "employed a place-value system and decimal notion using the nine 'Arabic' numerals and zero in his work." In the next century, Brahmagupta (b. 598 AD) estimated "the circumference of the earth that was one of the most accurate of any ancient astronomer." McLellan and Dorn also state, "In his *siddhãnta* Brahmagupta extended earlier work on measurement, algebra, trigonometry, negative numbers, and irrational numbers such as π. Indian mathematical work became known to the West primarily through reports by the eleventh-century Islamic scientist al-Bīrūnī in his *History of India*." In medicine, too, India was quite advanced compared to other parts of the world: "At their heights Indian medicine and surgery were probably the most developed and advanced of any contemporary civilization." They had medical schools and were quite knowledgeable about the parts of the body and how they work. By the twelfth century, medicine had become professionalized, and doctors taught their students at sophisticated schools.[7] There were blue-collar accomplishments as well. In particular, Indian accomplishments were most visible in the techniques of producing textiles, especially cotton and silk. Muslin and calicoes with excellent fade-resistant properties were produced during this period, and the superiority of the Indian fabrics was recognized worldwide into the 1700s.[8] These goods were produced mostly by hand; Indians made little use of machines during this period except for the spinning wheels that were used to manufacture luxury clothes and other items for royalty and nobility all over Asia. The innovation lay in the design of highly specialized tasks and in what we would call quality control today. The quality of Indian textiles was not matched by machine-produced goods until recently. Labor was so cheap that even with textiles of comparable quality, Indian goods were competitive for quite some time.

The Indian governments were very much in the mold of the old autocracies, with extensive bureaucracies that supported the kings and emperors. During the Gupta era, the caste system became highly structured, with the definition of about three thousand different hereditary castes.[9] Many of these castes defined and subdivided the trades and crafts into highly specific economic niches. A son born to a weaver became a weaver, and it was frowned upon, or even forbidden, for him to attempt to be anything else. Under such conditions, it is impressive that any blue-collar innovation took place at all. The workers were given subsistence wages at best and assigned highly circumscribed tasks. There was little incentive to increase the efficiency of the processes.

The Chinese used metals and developed techniques for identifying promising sites for mining. In the second century BC, Chinese metallurgists found ways of producing iron.[10] By the eleventh century AD, they were using water-powered bellows and smelting iron ore with coke, producing over one hundred thousand tons of iron annually, much of it for military use. To deal with frequent earthquakes, they invented an early version of the seismograph. According to Teresi, it was China that developed the four inventions that are generally considered to be among the most important in their implications for world history—namely, gunpowder, the compass, paper, and block printing. Gunpowder was invented in the ninth century. It was typically placed behind an arrow, stone, or other weapon in the cannon and ignited. As the gunpowder burned, it would launch the weapon. Crude compasses were used during the same period by travelers, and a process for magnetizing needles was developed.[11] Paper and the techniques for its manufacture were improved to the point that it was cheap and practical to make. Block printing was also developed there and was used to print books.

Other inventions included "cast iron, porcelain, sternpost rudders for ships, canal lock gates, stirrups and harnesses for horses, fishing reels, hot-air balloons, the seismograph, whiskey, gimbals, the umbrella, crank handles, kites, mechanical clocks, paper money, convertible bank notes, and many agricultural innovations, such as row cultivation, the iron plow, and the seed drill."[12] In fact, the acclaimed historian of Chinese science and technology, Joseph Needham, identified about 250 significant innovations prior to 1600.[13] While all of these would eventually be adopted in Europe, it was the adaptation of paper, stirrups, harnesses, compasses, mechanical clocks, and gunpowder that would have enormous consequences for the world. In the eleventh century, Chinese studies of time culminated in an astronomical clock designed by Su Sung. It was a huge and complex piece of engineering, housed in a forty-foot-high tower, and was powered by a waterwheel and complex gearing mechanism.[14] By the twelfth century, China had built fifty thousand miles of canals and waterways to support the transportation of goods and people. The great buildings, structures, waterways, and other projects demonstrate that the achievements of the Chinese autocracies exceeded those in the West.

China was the most populous society in the world throughout most of its existence, and its geographic area was huge. The Chinese achieved a stable autocracy headed by emperors that lasted for millennia. It was monolithic in that there were no independent governmental bodies or centers of power strong enough to challenge the central authorities. Diamond notes that one feature that made this central control possible was the geographic connectedness of the area. Military units could swiftly be sent out to any trouble areas, and communication between most parts of the empire was unhindered by mountains or oceans. The emperors claimed to have a "Mandate of Heaven," so while they weren't considered to be deities in the Egyptian mode, they still had the religious mystique. China was far too large and complex for a favored group of families to run the economy; instead, a huge bureaucracy evolved to support the emperors, numbering as many as one hundred thousand civil servants, not counting the military.[15] The sheer size of the population and the ruling class made a loyalty-based economy unfeasible, and a merit-based system was established. Examinations were given at the local and regional levels, as well as at what we would call the national level. Those who scored highest on the examinations moved into the bureaucracy and into the schools of learning. In principle, anyone could take the examination, but in practice it was mostly sons of the richer families who could obtain the necessary training in reading, writing, mathematics, and other subjects to get adequately prepared for the tests. The noble families had a large stake in getting their sons into the system and assuring their continued social status. The examinations at the national level determined who would have the most influence in the society, so it is easy to see that the whole system was not geared toward generating a wide variety of conflicting ideas. White-collar innovation was limited to contributions by the few who made it to the top of the heap. Blue-collar inventions probably happened either outside the bureaucracy or with the encouragement of local bureaucrats who were bright enough to recognize the usefulness of new ideas in carrying out the emperors' grand designs and enlisting the talents of craftsmen. The government owned all the resources and controlled all industries.[16] Workshops and factories were run by bureaucrats.

The Islamic Empire Becomes a Center of Innovation, AD 650–1100

In the seventh century, a new civilization emerged in the Middle East. Initially it was a small state in the Arabian Peninsula under the rule of the prophet Mohammed, but after a century of military invasions, it encompassed the current Middle East, Persia, North Africa, and Spain, all under the uniting religion of Islam. The Byzantine Empire was correspondingly reduced in size. The Islamic Empire was large enough that central control was not as pervasive as with previous empires, but the religious unity was strong enough to overcome the local rivalries. Except for those vying for power, many people were able to put their energies into commercial and even intellectual activities. Trade with India, China, and other countries expanded, resulting in a blossoming of commercial activity and interactions with other cultures. There was a tremendous interest in foreign writings of every kind, and books and manuscripts were translated into Arabic from ancient Greek, Babylonian, Persian, Indian, Egyptian, Chinese, and other languages.[17] Ideas in medicine, mathematics, literature, and philosophy were assimilated into the Islamic experience. Centers of learning sprang up in Baghdad, Damascus, Alexandria, and Cairo and later in Cordova, Toledo, and Sicily, with libraries that contained hundreds of thousands of scrolls and books. Whole towns were devoted to the manufacture of paper for books. For five hundred years, the Islamic world was the center of Western civilization; a significant number of its people were tolerant, innovative, and open to outside ideas. "Houses of Wisdom" were sponsored throughout the empire, acting as a magnet and attracting bright thinkers from Europe and the Middle East. Maria Menocal (2002) gives a wonderful description of medieval Spain, depicting a society governed by Islamic leaders, but open to new ideas and tolerant of other religions.

Islamic adventurers and traders developed a reputation as the best seamen and navigators in the world, and they were sought out by the Chinese and Indians and later by the Italians, Portuguese, and Spanish. The astrolabe, which had originally been invented in Alexandria, was developed into a useful instrument by the Arabs.[18] They also invented the quadrant, a simplified astrolabe that could be used to solve standard problems in spherical astronomy (i.e., locating stars, planets, and moons at different times from different places on earth). Astronomers used these instruments during this period to identify and locate hundreds of major stars in the sky, which is why they have the Arabic names that are in use today.

Among the great minds of the time was al-Bīrūnī (973–1048), who lived in Uzbekistan. He studied minerals in China and India, as well as in the Middle East. After visiting India, he concluded (correctly) that the Indus valley was an alluvial plain that had once been under the sea.[19] He wrote about the changes in the courses of rivers over a thousand-year period from Ptolemy's time. He also used triangulation to estimate the radius of the earth to be about 6340 kilometers, which is accurate to better than 1 percent (the mean radius of the earth is about 6371 kilometers).

Another great mind was the scholar Avicenna (Ibn-Sina, 980–1037), who lived in Persia and translated many of Aristotle's works. Using Aristotle's *Meteorologica* as his starting point, he wrote a book on the formation of rocks, in which he stated (accurately) that meteorites come from space and fall to earth, whereas "Aristotle claimed that they originated on the earth and were projected into the heavens by wind."[20] He classified rocks into four types, a scheme that was used in the West into the 1700s.

The Islamic scholars didn't just record knowledge; they added many innovations of their own in optics, astronomy, medicine, and mathematics. (Algebra was invented by the Persian mathematician Al-Khwārizmī around 820). Today we use their numbering system, including the concept of zero, which was developed while Europe was still using the awkward Roman numerals. To get a sense of the intellectual chasm at the time between the Middle East and Europe, consider that in 967 the monk Gerbert was accused of consorting with evil spirits for using a zero in his writings.[21] Fortunately for him this wasn't fatal, and he went on to become Pope Sylvester II.

Alhazen, whose Arabic name was Ibn-e-Haytham, conducted a number of experiments with lenses and mirrors, culminating in 1021 in his famous *Book of Optics* (in Arabic, *Kitab al-Manazir*), which was considered the definitive work on the subject for several hundred years. His experimental methods, which were scrupulously followed, earned him the reputation as the pioneer of the scientific method. He accurately explained and demonstrated the "camera obscura" and showed that the smaller the pinhole, the sharper the image. He developed the theory of vision that identified the lens-like quality of the eye.

Jabir ("Geber" in Europe, full name Jabir ibn Hayyan) was a chemist who developed experimental equipment for testing chemical reactions. He emphasized the importance of experimentation above speculation, and he is considered by many to be the father of chemistry. He documented the differences between acids and alkalis and identified a number of substances with these properties.

There were many other Islamic scientists and mathematicians who made significant contributions. According to McLellan and Dorn,[22] Islamic civilization was the world leader in all fields of science from about 800 to 1300.

The Islamic Empire produced a number of blue-collar inventions as well, including crankshaft-driven windmills, water turbines, and the most advanced mechanical clocks.[22] It is also credited with the invention of the spinning wheel.

The Islamic Empire consisted of several societies united by a common religion. All were autocracies in the classic mold. Possibly because there was some competition between them, the rulers established Houses of Wisdom and supported scholars to demonstrate their own wisdom. Scholars, being naturally curious, looked far and wide for ideas, and this paid off for several hundred years.

Unfortunately for the societies within the Islamic Empire, and indeed for the rest of the world, white-collar innovation fell off significantly after about 1300. This was also true of the blue-collar innovation that we know about. This phenomenon has been lamented by Islamic scholars and argued at length by Western writers (see Mernissi (1992), Lewis (2002), and Landes,[24] to name a few). Their analyses go into great detail about the effects of religion, cultural values, and the attitudes of the rulers. What seems to be a common theme is that the very success of the Islamic Empire led to a kind of cultural arrogance of "we're the greatest" and a tendency to denigrate the ideas and inventions of outside societies. Obviously the Islamic Empire was far from being the only society to fall into this trap.

Blue-Collar Innovations of Europe in the Middle Ages, AD 900–1400

While the center of much white-collar innovation was in the Islamic societies, blue-collar innovation was stirring in Europe, under conditions very different from the earlier autocracies. This would have enormous consequences, as we shall see.

Virtually no invention arrives on the scene in its full embodiment. The first form of an invention offers a new capability, but it usually has severe limitations. However, contained in its concept is the prospect of improvements (e.g., making it stronger, more powerful, cheaper, lighter, bigger, smaller, easier to build, or faster or adapting it to a different environment). Once the initial capability of an invention is demonstrated, clever people endeavor to fully understand its workings, materials, and fabrication, and then they come up with ways to extend its capability.

For example, once it was demonstrated that the flow of water could cause a wheel with paddles to rotate, innovative people realized that there were a number of possible uses and improvements:

1. A stream could be used instead of people or animals to move heavy objects.
2. The turning axle could be used to turn millstones for grinding grains or wood for paper, to lift heavy weights with ropes, or to lift a loop of water buckets for draining a mine.
3. The turning axle could be connected to gears that would change the direction of rotation that would make it more convenient for some uses.
4. The stream could be channeled to make the waterwheel more efficient.
5. This channeling could be controlled in order to stop the motion of the wheel or control its speed.
6. Water could be channeled to come from above ("overshot"), rather than under the wheel ("undershot"); this would increase its power.
7. Better wheel materials, in this case different woods, would last longer in water.
8. Better techniques for lubricating the axles would extend the wheel's life.

These and many other improvements were made to realize the potential of water as a source of energy. When inventive people were at work, they were trying to solve specific problems that had immediate payoffs. Their concern was not the long-term implication of what they were doing. However, the long-term effect was considerable; namely, they increasingly used available natural resources to replace humans and animals as sources of power.

Waterwheels were built all over Europe. In England alone, there were about 5,600 mills that used water wheels by 1086.[25] In Europe overall, there were on the order of half a million waterwheels at that time.

What distinguished the European experience from that of China during this period was not that there weren't a similar number of initial embodiments of inventions, but that in Europe there were more opportunities to explore variations, improvements, and adaptations. Consequently, the long-term effects were significantly greater—they resulted in significant changes in the ways the societies functioned.

As with the waterwheel, Europeans adapted the windmill to local conditions and available materials. Ancient windmills were horizontal and rotated on a vertical shaft. This had the advantage that it didn't matter from which direction the wind came, but because the wind pushed to some extent on both sides of the vanes, the windmills were not very powerful. Early windmills in England were vertical but did not rotate, which limited their use to those times of year when the winds were coming from the proper direction. The primary use of windmills was drainage, so it was desirable to be able to harness the wind whenever it blew significantly. Innovative individuals then came up with the post mill, which could be rotated by hand so that a human could adjust the blades to face the wind regardless of the direction. Later, the tower mill, in which only the top cap rotated, was developed; this made it possible to drive machinery within the tower, although this design was more expensive. The windmill enabled Holland to exist as a functioning society because vast areas were drained using these machines and made available for agriculture and housing.

The waterwheel and windmill transformed Europe by harnessing the power of natural forces to replace humans and animals. Several other society-changing innovations have been cited by historians—namely, stirrups, gunpowder, pulleys, levers, double-entry bookkeeping, the compass, eyeglasses, the mechanical clock, and the printing press.

Stirrups, like gunpowder and the compass, were invented in China, not Europe, but warriors working for feudal lords recognized their usefulness: they provided a lot more stability for horseback riders. This enabled mounted knights to carry and wield heavier weapons and to swing and thrust heavy swords without falling. The knights became much more fierce and effective and played a key role in replacing infantries with cavalries in military confrontations.

Gunpowder and the cannon were invented by the Chinese before AD 1000. According to Landes,[26] the Europeans and the Islamic world probably learned of them in about 1300. By the sixteenth century, the Europeans developed means of compressing gunpowder into kernels that made for stronger explosions, and because of the air pockets between the kernels, the burning was much faster, making it launch the weapons farther and enabling the use of heavier stone or iron balls that no fortress could withstand. Wheels were added that made it possible to orient the cannons in any directions and move them from place to place. While the early embodiments of the cannon put the European armies on a par with their Middle Eastern adversaries, the adaptations and improvements that followed gave them a decisive advantage in both offensive and defensive actions. The short-term advantage was to improve the performance of individual military units, but the long-term implication was that it helped establish Europe's military dominance in the world.

Levers, pulleys, and other similar devices proliferated in Europe. Block-and-tackle pulley systems, which date back at least to Archimedes in the third century BC, were used to lift heavy objects like stones, buckets of water, and building materials. In Europe during the Middle Ages, they were used in the construction of fortresses, castles, cathedrals, palaces, and public buildings. Evidence of their use can be found throughout Europe. At a national gallery in Urbino, Italy,[27] there is an impressive display of plates that illustrate a wide variety of mechanical contraptions used in construction from 1400 to 1700: winches, pulleys, waterwheels, gears, pumps, weights, and clocks. One plate shows how an obelisk was raised from a horizontal to a vertical position; another shows a drawbridge, with teeth on a wheel that moves the bridge sideways. Other examples of the complexity and variety of construction devices can be seen at a French abbey,[28] where drawings on display show concentric wheels of different diameters using tied-down ropes. There are many more of these displays throughout Europe.

In addition to the societal impact of waterwheels and windmills, historian Landes places considerable importance on the development of eyeglasses and the mechanical clock.[29] While shaped glass had been used earlier for magnification, it wasn't until about 1300 in Italy when spectacles were invented where the glass was mounted on a frame that could be worn, thus freeing both hands for useful tasks. While the immediate advantage was to improve the visibility of small objects and writing, the invention had several world-changing effects: (1) it greatly extended the effective work life of many craftsmen who otherwise couldn't see well enough to do good work as they got older, (2) it enabled scholars and other literate people to read throughout their lives, and (3) with magnifying eyeglasses, it created the potential for the development of fine instruments that could be made smaller and fabricated more accurately and conveniently. As with every important invention, the initial designs were subsequently improved upon; in the next few centuries, spectacles of different shapes, both convex and concave, were developed to accommodate a range of visual problems involving farsightedness and nearsightedness.

The compass was invented in China and was adapted in Italy in the late thirteenth century to operate in a shipboard environment, with its pitching and rolling.[30] The Venetians recognized the capability of the ship borne compass and used it in a novel way. With these instruments on their ships, they were no longer limited to sailing during the summer, because they could navigate even in cloudy weather. The compass provided the direction of the ship, and they could estimate the ship's speed in the water by estimating the length of time it took for an object dropped at the bow to pass the stern. This technique was accurate enough in the Mediterranean between Venice and the Middle East to ensure safe travel. It also enabled the Venetians to build much larger ships that were sturdier, carried a larger cargo, and could be operated with fewer mariners. The utility of the compass was greatest on well-established trade routes, where the mariners had prior knowledge of the geography and the sea depths. Later, when combined with the sextant that provided latitude information, it enabled mariners to travel into uncharted waters and still find their way back to port.

Inventing a technique for timekeeping has been the subject of innovative activity since antiquity. Sundials were in use since ancient times, but of course they don't work at night or on cloudy days, and they are not good for measuring time to minutes or seconds. Water clocks were invented and improved in China, Greece, Rome, the Middle East, and Europe. They used a controlled flow of water to turn a waterwheel, which then was connected in a number of ingenious ways to display time and the passage of time. The first mechanical clocks that did not use water were developed around the thirteenth century in Europe. They were prone to breakage and required constant attention, but they worked. They not only provided a visual display for time but also could be rigged up to ring bells at specific times, such as on the hour, leading to the use of hours, minutes, and eventually seconds. Improvements over the next few centuries included more durable materials, more accurate escapements (oscillating components that regulate the ticks of the clocks), spring mechanisms to provide the driving energy, use of a pendulum to increase accuracy, and smaller components that ultimately resulted in watches. Whereas the early water clocks were used mostly as toys and gifts for the amusement of the royalty, the European clocks were installed by local authorities in towns and villages, and this ultimately changed the way people lived. All village dwellers knew when it was time for meals and how much time they worked. They were able to schedule events for a particular time, and people knew when to show up. People were able to utilize their time more efficiently for work and pleasure.

During the period from the fall of Rome to about 1500, several developments took place in Europe that prepared the way for innovative activity. Why they had such an effect will be explored in later chapters. After the breakup of the Roman Empire, political control was in the hands of a large number of feudal lords and warrior tribal leaders. Eventually the feudal lords banded together under monarchs. These monarchs fought each other, negotiated with the feudal lords, and attempted to re-create the old autocratic conditions. However, they were not successful, partly because none became powerful enough to establish dominance over very much of Europe. They also had to share power with the feudal lords and with the emerging Church, which eventually took over the governing of the people's religious and spiritual lives. The Church went along the "divine right of kings" as the notion further legitimized its spiritual authority. The monarchs accepted the idea because, while they couldn't claim to be deities or quasi-deities, it did associate their authority with God and served to intimidate their subjects.

The Church became a more centralized and powerful political body, with its own courts, armies, and land. With its independent access to funding, it built great cathedrals; in the process, it encouraged bidding by different architects, contractors, craftsmen, and artists, who competed for the contracts. The Church also encouraged the establishment of monastic estates, which required considerable labor to operate. Since these activities were outside the old autocratic monopolies, there was less motivation to keep innovations secret, so a number of inventions were shared across the continent between the monks, especially among the Cistercians. The cathedrals were constructed in villages and encouraged the growth of towns.

While many peasants were bound to the feudal lords as a legacy of laws promulgated late in the Roman Empire, these bonds were more easily broken as peasants found ways of moving from one estate to another and into the growing towns. The plagues accelerated this trend. The towns needed governance, and their leaders successfully played off the feudal lords against the monarchs and achieved some independence from both, and this independence increased with time as the towns grew in size and number. The economic activity that fueled this process was increased trade between towns, between kingdoms, and with distant empires and kingdoms. Trade required the production of goods, both raw materials and finished goods. This trade was made possible in part by the surviving

Roman roads and in part by the wide-area trade centered in the city-states along the Mediterranean. The Venetian Republic was the most influential, but Genoa, Florence, Milan, Amalfi, and Pisa were also significant.

Venice was one of the first plutocracies[31] and was run by several hundred families who selected the heads of government. The following description of how the Venetian Republic evolved is taken from Acemoglu and Robinson.[32]. Shortly after it gained independence, Venice established a practice called *commenda*, which was essentially a temporary partnership between a financier and an adventurer, with the goal of making money from trade. The financier supplied most of the upfront investment money and was usually a member of a wealthy family; the adventurer was typically an ambitious commoner who developed the navigational and managerial skills necessary to purchase and deliver the goods via ship back to Venice. The two partners then shared the profits. This practice enabled some talented and lucky adventurers to become quite wealthy and to marry into and join the aristocracy. This class mobility and the trade opportunities in the Mediterranean and beyond fueled an economic expansion that was shared widely enough among the classes that the Venetians achieved the highest level of prosperity ever experienced in the world up to that time. Its population grew to 110,000 by 1330, making it one of the largest cities in the world.

Unfortunately, this desirable state of affairs did not last. The old-money families were threatened by the newly rich and by the early 1300s put an end to the *commenda* practice and consolidated their hold on the economic and political life of the republic. While trade continued, Venice stopped growing in population and prosperity, and its innovative activity declined.

During the Middle Ages, there was little interaction between white-collar and blue-collar innovation. According to McClellan and Dorn,[33] "The impressive array of technological innovations that led to the transformation of European society and culture owed nothing to theoretical science, in large measure because science had little to offer." This situation was about to change.

5. Innovation Gathers Steam

Innovative activity picks up in Europe;
some societies become distinctively more innovative than others;
blue- and white-collar efforts begin to cross-fertilize

In the four centuries leading up to 1800, populations in Europe increased, towns and cities were established and grew, kingdoms became nations, international trade increased, nations conquered and exploited the Americas and other parts of the world, merchants gained power and wealth, and mercantilism became widely practiced. While a few prospered greatly, the result was a modest improvement in most people's lives in Europe and some colonies. In other parts of the world, populations and trade increased, but personal income didn't change significantly. *The worldwide personal well-being, as measured by the annual GDP per capita, remained virtually unchanged at about GK$400–GK$600* from the beginning of civilization until the 1800s.*[1] With the growth in population, there were more rich people, but there were correspondingly more poor people; the middle class grew significantly only in a few places like the Netherlands and other western European countries.

The rise of Protestantism encouraged challenges to Catholic thought and opened the way for white-collar innovation to thrive, and thrive it did during this period. Curious aristocrats, both Catholic and Protestant, began to ask questions about nature in a new atmosphere of freedom, even daring to dispute the ancient geniuses.

Trade between kingdoms played a central role in the changing environment for innovation. Between 1000 and 1600, trade expanded significantly as the world population doubled, from about 270 million to about 560 million people. The world GDP almost tripled during this same period, so it is reasonable to conclude that trade at least doubled during this time. After the Venetian Marco Polo visited China late in the thirteenth century, trade between the Italian city-states and China increased. The Venetian Republic functioned as a plutocracy for several hundred years.[2]

---

* *Here and throughout the book, the symbol GK$ indicates Geary-Khamis dollars as defined in reference 1, and all figures involving GDP, population, and GDP per capita are obtained from this reference. Roughly speaking, a Geary-Khamis dollar has the same value as a US dollar in the year 2000.*

The Venetian Republic established institutions that made the emergence of an autocrat virtually impossible, whereas the Roman Empire had not done so. While the doge was officially the head of the government, he was elected by a group of powerful families, and his power decreased over time. During the life of the republic, this group of families divided up the functions and powers into a number of government bodies (like the Signoria and the Council of Ten) that distributed the executive, legislative, and judicial duties and authorities. The families of power were well aware of the dangers of excessive concentration of power—all they had to do was look at the governments they traded with. While they wanted to have the benefits of great buildings and other shows of wealth, they didn't want the violence and the limits on their freedoms they saw in the other cities. They were almost obsessive about it; for example, in 1454 they established a supreme tribunal that used spies and surveillance to enforce this policy. Along with these institutions went a cultural conformity that included a dress code. When appearing in public, some government officials were required to wear certain identifying colors. This can be seen in portraits of Venetian political leaders, painted by Renaissance artists like Titian.

With the growing trade, new jobs were created in the city-states throughout what is now Italy, so more people possessed enough money to more than cover their subsistence requirements. This meant that there was a growing market for domestic things like clothes, fancier foods like spices and teas, furniture, and other household goods. The annual GDP per capita rose from the worldwide average of GK$450 in 1000 to GK$1100 in 1500, reflecting this trend. The Italian city-states were the richest societies in the world. Interestingly, from 1500 the GDP per capita did not increase further for the next three hundred years even though the population doubled, which indicates that the early growth spurred by the *commenda* in the Venetian Republic was not sustained by the more closed plutocracies there or in the other city-states of Italy. The increase in the market just kept pace with population growth.

The rest of Europe was a different story. From 1500 to 1800, the annual per capita rose by about 50 percent, from just under GK$800 to GK$1200, ultimately surpassing the Italian city-states. Since this period coincided with the colonial conquests of the Europeans and their adoption of mercantilist practices, most of this increase probably came from the exploitation of their colonies, whose economies remained just above the historical subsistence level. Regardless of its source, this increased prosperity created larger markets for domestic goods in

Europe. People of all classes wore better clothes of wool and cotton, so the markets boomed for woven goods and many other items.

The Fifteenth and Sixteenth Centuries

Following the lead of the Italian city-states, the ruling classes throughout Europe found it in their interests to establish universities, become literate, and sponsor scholars and natural philosophers, congratulating themselves for being enlightened and open to new ideas. Intellectual activity was encouraged if it didn't contradict Church doctrine, and even this limitation was weakened by the Protestant Reformation and the birth of new denominations that didn't consider the Vatican to be the final authority of religious, philosophical, and historical ideas. While there was considerable intellectual activity in clerical circles, there was not a lot of white-collar innovation concerned with explanations of the natural world. One exception was Copernicus, who theorized in his book *De Revolutionibus* (1543) that the model of a solar system, with planets including the earth orbiting around the sun, was much simpler and more accurate than the previously accepted Ptolemaic epicycles. Another was Andreas Vesalius, who wrote the book *De Humani Corporis Fabrica* (1543) that described his theories about the workings of the human body based on actual dissection.

Blue-collar innovations continued to grow, improving on the waterwheels, windmills, eyeglasses, compasses, and mechanical clocks. Two major innovations of this period stand out: double-entry bookkeeping and the printing press.

The growth of commerce in the Mediterranean and between the city-states of Italy created the need for standardized accounting methods. Embezzlement by employees was also a problem, so over the years a number of different accounting methods were developed in Milan, Venice, Genoa, and Florence. Agreements between parties were recorded, along with the prices and names of the parties and witnesses. Different types of books were kept, such as journals and ledgers, and the concepts of credits and debits were used. By the end of the 1400s, the concept of double-entry bookkeeping had been developed, wherein for each transaction one account is debited and another credited. The accounts were labeled as assets, liabilities, and owner equity, and the framework was developed so that assets = liabilities + equity. While initially this method was employed to provide a record of transactions and a means of estimating how well the business was going (by looking at the equity), it eventually became evident that it also helped reduce employee theft and provided a fairly rigorous way to detect and fix mistakes, which were (and are to this day) numerous. An interesting fact about the bookkeeping methods is that Roman numerals were used for several hundred years before they were gradually replaced in the 1500s by the Arabic system we use today. Because of the association of the Arabic system with Islam, their use was frowned upon in Christendom, but businessmen increasingly performed their computations using Arabic numerals, which was far easier. One might question whether this innovation should be considered as blue-collar. According to the definition used in this book, it is indeed, because double-entry bookkeeping is utilitarian; the computation involved does not involve complex mathematics, just lots of simple arithmetic. Employees of the merchants performed the day-to-day bookkeeping. So the blue-collar association still fits, even though we generally associate bookkeeping with offices and white-collar employees.

The printing press was a revolutionary invention and is considered by many historians to be the most influential one of all time. As revolutionary as it was, like every invention it was solidly based on prior developments. There were already inks and papers with the properties necessary to leave a clean, consistent image of a character on the paper after being pressed against a typeface with ink. These had already been established with the woodblock printing techniques developed in China several centuries before. Movable type, wherein large numbers of small blocks could be pressed together in bunches to form a page of text, had also been developed earlier in China. However, Johannes Gutenberg (circa 1395–1468) recognized that these techniques could be vastly improved to make it cheaper and produce readable text. As a goldsmith, he understood how metals could be melted and formed, so he realized that if he could come up with some way to quickly replicate the "types" containing characters and find a means to quickly apply ink to the face of the types and press them against paper, then he could reproduce printed pages cheaply and rapidly. To make this all happen, he invented the hand mold, whereby a single set of punches in the shape of letters could be used to create a mold from which types could be formed by repeatedly casting molten metal. He also saw that he could adapt the screw press, used for pressing wine and olives, to stamp out pages rapidly and consistently. His earliest printed documents were small papal indulgence documents, but they were soon followed by a complete Bible in about 1455. By 1500 something like two million books had been printed, and by 1600 this rose to over a hundred million, mostly in Europe.

Blue-collar innovation got a big boost with the invention of the printing press, although it would be centuries before its effects were fully felt. No longer limited by the constraints of oral communication or hand-copied instructions, the printed word eventually included observations and instructions for all manner of enterprises and pursuits. Less appreciated is the fact that the printing press also enabled the mechanical reproduction of pictures. One of the first publications that used print to communicate technological methods to succeeding generations was the book *De Re Metallica* by Georgius Agricola (this was his scholarly Latin name—his German name was Georg Bauer). Published in 1556 after being in preparation for twenty-five years, it described in great detail the machines, techniques, processes, and materials using in mining and metallurgy at the time and included surveying methods and even legal and safety measures. It included hundreds of plates that made it a classic reference for at least two centuries. While any one of the specific tools or procedures in the book could have been described orally, the sheer number of small innovations and their interdependencies would have been impossible to communicate accurately solely by word of mouth and hand-drawn pictures. No doubt some people weren't happy that Agricola made all this information widely available and gave away many secrets of the trade. The book did, however, set a precedent whereby experts might gain recognition and hopefully some income by publishing. It gave successive generations access to valuable information and provided guidance in the application of technical knowledge.

While the printing press opened the door to the distribution of knowledge to everyone, centuries would pass before the effects of the printing press were widely felt among craftsmen. It didn't happen until their literacy was more widespread, thus enabling them to make use of the collective knowledge of their predecessors. When this happened, the distinction between blue-collar and white-collar innovation virtually disappeared in many subject areas. Plus, a higher percentage of people in the technologically advanced societies became literate and knowledgeable, greatly expanding the base of potential innovators.

The Seventeenth Century

Up to the seventeenth century, white-collar and blue-collar innovators operated in separate worlds, with little interaction between the participants. The thinkers had little impact on the inventors, who used trial-and-error, rules-of-thumb, elementary arithmetic, and lots of common sense in their efforts. The resulting inventions, in turn, did not cause significant shifts in the thought patterns of the thinkers. This would change: the next few hundred years would witness a merging of white-collar and blue-collar innovation, a marriage of science and technology.

Early in this period Galileo Galilei confirmed the Copernican hypothesis using a telescope that he had made. He also developed a procedure by which longitude could be determined through observations of Jupiter's moons. There being a community of scholars in Europe (and a mail system), the findings of the experimenters and theorists were quickly communicated among peers, making it very difficult to prevent ideas from being spread. Johannes Kepler developed the orbital equations of the planets, Isaac Newton formulated the law of gravitational attraction, and a number of scientists established the equivalency of several different types of energy and how they could be exchanged. Christiaan Huygens developed a theory of light propagation, which asserted that light emanates from its source in waves, not as particles as Newton had claimed.* The microscope was invented during this time, and Antonie van Leeuwenhoek used it in 1676 to become the first person to observe bacteria directly. Robert Hooke and Robert Boyle studied the properties of gases and formulated fundamental laws about their behavior, which would prove to be crucial in the development of the steam engine. Boyle also developed an atomic theory of matter, which helped establish the scientific formulation of chemistry and distinguished it clearly from alchemy.

In the blue-collar world, innovations were primarily (1) improvements to existing inventions like waterwheels, windmills, lenses, and the printing press and (2) improvements in tools, metalworking techniques, mining machinery, and accounting techniques. During this period Galileo and others developed the telescope, the thermometer, the barometer, and surveying tools, which were some of the early cases where the blue- and white-collar distinction was reduced. These devices proved to have utility in other areas than scientific experiments. For example, telescopes had military applications and enabled people in port cities to spot and identify approaching ships much sooner than before, and thermometers and barometers provided a means of quantitatively recording temperature and pressure, which were found to be useful both in scientific experiments and in other areas of life.

---

* *The wave theory of light was eventually accepted as the superior model. However, in the twentieth century all electromagnetic radiation, including light, was shown to have qualities both of waves and of particles.*

While a few natural philosophers (who would be later known as scientists) were exploring exciting new theories, and a few "low mechanicks"[3] (who would be later known as engineers) were making improvements on machines, Britain's monarchs continued to lose political power, to the benefit of the world. The Magna Carta, which was imposed in 1215 on King John of England by a group of feudal barons, limited the power of the monarchy and protected the barons' privileges. Its impact was limited until the 1500s, after it had been printed and widely distributed. As merchants acquired wealth through the growing trade, they also claimed the rights proclaimed in the document, and a sense of entitlement to certain freedoms began to spread.

By the seventeenth century some societies were showing economic progress. Although their lives were improving a little, the people at the bottom of the economic ladder were still quite poor, and the rich classes reaped most of the benefits from the improved trade. In many parts of the world, including most of Europe, increased economic activity was offset by population growth, the result being that poorer individuals were not significantly better off economically. The GDP per capita of western Europe increased somewhat from 1500 to 1700, rising from about GK$800 to about GK$1000. One exception was in the Netherlands, which established itself as the major colonial power in the world, dominating worldwide trade and bringing wealth back to the mother country. While the average GDP per capita of the Italian city-states leveled off between 1500 and 1800, the Dutch income rose to GK$800 in 1500, to GK$1400 in 1600, and to over GK$2,000 in 1700, where it leveled off as the Dutch colonial era came to an end. For over two hundred years, the Dutch were the most prosperous people in the world.

With books being printed everywhere, more and more people began to recognize the value of being able to read and write. As mentioned in the previous chapter, the trading cities of the Mediterranean needed larger numbers of people to keep accounts, manage the increased volume of commercial transactions, and communicate with people from other countries. With this trade came the beginnings of a middle class in the participating towns and cities.

Outside Europe, there was no such improvement. While populations grew, production grew in lockstep, indicating that the additional wealth accrued to the aristocrats and a favored few. While the rich there grew in numbers, so did the poor, and the plight of the poor didn't improve. This situation was compounded by colonialism: the European countries treated their colonies like feudal estates and as sources of natural resources like metals and crops. Throughout Asia, which included many of the colonies, the GDP per capita remained at about GK$600 during this period.

The Eighteenth Century

During the eighteenth century most white-collar innovators were mathematicians and experimental scientists; they were amateurs in the sense that they were aristocrats and did not get income from their efforts. Prominent among the mathematicians were the Swiss Daniel Bernoulli and Leonhard Euler and the French Adrien-Marie Legendre, Joseph Lagrange, and Pierre Laplace. One famous experimental scientist of the time was the French chemist Antoine Lavoisier, who is considered to be the father of modern chemistry. He formulated the concept of the conservation of mass during chemical reactions and made a clear distinction between elements and compounds. Less well-known was the Spanish scientist Antonio de Ulloa,[4] who is credited with discovering the element platinum. He went to Ecuador on an expedition sponsored by the French Academy of Sciences to resolve a scientific dispute about the shape of the earth near the equator. During this century there were significant advances in the development of differential equations and the properties of fluids, which would later be used in practical applications. One interesting story involves Euler,[5] who used the new mathematical techniques to compute the optimum size and orientation of the sails on a Dutch windmill, an impressive feat. He found that the craftsmen, who over a long period of time had tried different combinations, had indeed ended up with designs that were close to the theoretical optimum.

While the white-collar innovations of the eighteenth century may have been less impressive to laymen than the earlier formulations of the law of gravity and the validation of the Copernican theory, the developments of mathematical tools to explain physical processes and the determination of the properties of materials were recognized as very important within the growing scientific community. They would also prove to be crucial to the development of important innovations of the next century.

It was in this century that scientists became interested in determining the properties of nature with an eye to developing useful instruments and machines, thus blurring the distinction between blue- and white-collar innovations. Benjamin Franklin, for example, studied the electrical properties of storms and invented the lightning rod. Instead of pursuing a patent in Britain or building a factory to make the rods, he was content to publish his design and let the public use it. Later he invented bifocals.

In the blue-collar world, one of the major areas of innovation involved practical uses of steam power. Thomas Newcomen, a literate ironmonger and blacksmith, invented a practical means of lifting water from a deep mine using a seesawing beam and a steam-driven piston.[6] Significantly, he was not funded by some aristocrat, but by a secret religious community bank (the first venture capital firm?). He was in communication with the scientist Robert Hooke, who had studied the properties of gases. Later James Watt developed efficient new designs for steam engines.

John Harrison (Sobel, *Longitude*) invented a series of increasingly small, accurate clocks that could work in a shipboard environment, culminating in the chronometer, a major innovation that enabled ships to accurately determine their longitude even after months at sea. In France, Claude de Jouffroy was the first to use the steam engine to propel a boat using paddles. John Kay invented the flying shuttle, a major advance in weaving that increased the productivity of weavers. Andrew Meikle invented the threshing machine, which mechanized the separation of grain from stalks and husks.

During the eighteenth century the blue- and white-collar worlds had some crossovers, as noted above, but in general they developed separately, as they had throughout history. Blue-collar developments became increasingly associated with technology, and white-collar developments became increasingly associated with science. Previously, most blue-collar inventors were illiterate and were limited by the knowledge gained from listening to their mentors, studying their mentors' drawings, and trying out a few things with restricted resources. What we now call technology requires access to information and training not available to illiterate people. In the white-collar world, the emergence of the scientific method reduced the amount of intellectual dabbling and required more rigorous experiments and observations to back up any theoretical claims.

Technological advancement was enhanced by literacy and training, which all the inventors mentioned above had. Craftsmen increasingly used printed documents to record their techniques and procedures. According to Rosen,[7] there emerged during this period in Britain an artisan class that was literate, trained in practical mathematics and engineering, and highly motivated to improve on each other's inventions. Inventions increasingly required detailed knowledge of the properties of materials like metals, coal, and water. Inventors had to consider the cost and ease of fabricating their products, the availability and cost of materials, the durability of their products, and the patentability of their inventions. Illiterate inventors were at a severe disadvantage.

"Natural philosophy" was the term most commonly used to describe the study of the physical world at this time. Benjamin Franklin founded the American Philosophical Society dedicated to scientific pursuits, not what we think of as philosophy. The term "science" became increasingly used to describe a discipline that involved a methodology of developing theories (i.e., models of reality) and devising experiments to test the theories against observed behavior. Between competing theories, the one that matched reality under a wider set of conditions gained acceptance as the dominant theory. To take an example from astronomy, a simplistic theory of planetary orbits modeled the orbits as circles around the sun. Observations of the planets showed that they came pretty close to being circular. However, measurements of the planets' locations with time were accurate enough to indicate clear deviations from circular orbits. Enter Johannes Kepler with his theory that the orbits are elliptical, just slightly noncircular. These same measurements showed that the elliptical model is much better, but still there were small deviations. With Isaac Newton's theory of gravitation, elliptical models are almost perfect for a single sun with a single planet and no moons. However, his theory enabled the computation (not simple) of the effects of moons and other planets on the orbits, and with these taken into account, the agreement between observations and theory became remarkably close.

So practitioners of science began to think of themselves as "scientists," and Galileo and Newton are considered to be two of the earliest examples. Their successes and those of other scientists in relating observations of the natural world to a new concept of the universe convinced most thinkers to discard the idea that the physical world could be explained purely by theorizing. The natural world was full of surprises that could only be discovered by experimentation. A story that illustrates the folly of theorizing tells of a philosopher who wrote that if you move a ball in a circle on a flat surface and then let go, the ball will continue to move in a circle—the hypothesis being that the ball will be trained. Of course, anyone who tries this finds that the ball actually moves in a straight line; if the writer had bothered to try it himself at his desk, he could have saved himself from looking foolish.

During this century international commerce increased to keep up with the growing populations in Asia and Europe. In Europe the monarchies continued to weaken, with the populations gaining a few more rights and becoming more literate. The United States was established as an independent country based on democratic principles; a presidency was established with significantly less power than the monarchies of Europe, along with legislative and judicial branches with the stated intent of establishing checks and balances on the distribution of power. Since power worldwide remained firmly in the hands of aristocracies, the plight of the average person improved only slightly, and then only in a few places. The worldwide GDP per capita went from GK$600 in 1700 to about GK$650 in 1800, most of the increase happening in Europe, where it rose from about GK$1000 to GK$1200.

6. Innovation Takes Off [1]

GDP per capita rises in innovative societies; capitalism shows its ability to create wealth; patents promote innovation; everyone loves innovators; innovation becomes institutionalized

Something extraordinary happened in the 1800s in Europe and North America in particular. A large number of people became significantly better off economically, lived longer, ate better, had healthier lives and more leisure time, and became taller and better educated.[2] Innovative activity of all kinds increased sharply. Scientific experiments proliferated, the number of inventions exploded, and societal attitudes of all classes toward innovation changed from wariness to a rush of enthusiastic acceptance. Inventors like Thomas Edison were folk heroes. It seemed like everyone embraced innovation.

It didn't happen over millennia or centuries, but over mere decades. It was pointed out earlier that worldwide personal well-being remained virtually unchanged from the beginning of civilization until the 1800s. This means that even while the world population increased, from about four hundred million in 1500 to one billion in 1820, the production and income of the typical family did not improve in most places in the world. In fact, in China, India, and most of Asia and Africa, the plight of the typical family didn't improve significantly until after 1950. However, in western Europe and in North America, annual average GDP per capita began growing before 1800, and in the 1800s it accelerated. The rest of Europe and the Americas experienced more modest improvements. After 1920 many other countries experienced impressive economic progress, growing incomes, and improved living conditions.

It is a fascinating story. Here is how it all happened.

The Nineteenth Century

In the nineteenth century, revolutionary changes occurred in which innovation played a dominant role. Words and phrases like "revolutionary," "remarkable," "earth-shattering," "quantum leap," and "unprecedented" seem inadequate to describe the changes that took place. During this century, monarchic powers were drastically reduced and balanced by popularly elected parliaments. People in the fortunate countries grew taller, healthier, and more prosperous. People's lifetimes doubled, literacy rates soared, and universal education became a reality for all classes of people. Food production rose far above the Malthusian trap,[3] so that population growth could be, and was, quite large. The world population went from about a billion people in 1820 to one and a half billion by 1900. In western Europe, the GDP per capita increased from about GK$1200 to GK$3,000. Given that for the ten thousand years up to AD 1000, the same figure for all societies in the world was about GK$400, this was remarkable. Equally remarkable, these trends continued, with some starts and fits, throughout the following century and spread to many other countries.

By 1800 the properties of water and steam were understood well enough to make practical machines. The Watt steam-engine design, the materials used, and the fabrication methods of the parts had all been improved to the point that steam engines could be readily built and produced for a reasonable cost. They had the desirable feature that they could supply a constant amount of power continuously, whereas waterwheels and windmills were at the mercy of nature. Animals and humans could not be harnessed to supply a comparable amount of power. Once this capability was recognized, it changed how all manner of goods were manufactured. Factories with a centralized source of power could produce in one place goods that previously were produced by individuals and small shops dispersed around the towns and countryside. The quality of the goods was higher and more consistent, and the products were more reliable and less expensive. Plus, each factory could produce larger numbers of products to meet the growing markets.[4] Rosen considers the steam engine to be the most powerful idea in the world, a claim difficult to refute. It was central to the Industrial Revolution.

During the nineteenth century, steam engines were used to propel boats, locomotives, and vehicles. Steamboats were used along rivers and between ports, and eventually steamships were used on the high seas. Steam engines were used to pull trains, revolutionizing the transport of goods on land. Inventors came up with steam shovels and steam-powered tractors and cars. Later in the century, steam turbines (which operate on a different principle) were used to produce electricity, and continue to do so today.

Before the century was out, inventors all over Europe were trying out ways to substitute the controlled explosions of ignited gases or vapors for steam, and eventually they developed a number of embodiments of the internal combustion engine that were practical and could be manufactured using available methods and materials. The new engines replaced steam engines in a number of applications. They had the advantage that they could be made much smaller than steam engines, and they didn't need the large boilers and coal- or wood-powered furnaces, just a container to hold the fuel. This feature was very desirable for vehicles.

The most dramatic innovations in the nineteenth century centered on electricity and the transmission of information. These developments required close cooperation between scientists and engineers, a classic marriage of blue-collar and white-collar innovation, between technology and science. Michael Faraday was born into a blue-collar family and never was formally educated, yet he became an outstanding scientist and was elected to the Royal Society at the age of thirty-three. There were few places in the world at that time where this would have been possible. His experiments in electricity and magnetism showed the close relationship between the phenomena and provided the experimental evidence for James Maxwell to formulate his famous equations that both identified the relevant parameters of electricity and magnetism and illustrated the relationships between them. Maxwell concluded that light had both electric and magnetic qualities and was a form of electromagnetic radiation. Heinrich Hertz demonstrated that radio waves had the same properties, except that the frequency of the radiation was much lower; radio waves also traveled at the speed of light. Later it would be shown that infrared, ultraviolet, and gamma rays were also forms of electromagnetic radiation, all at different frequencies.

Faraday's experiments showed that electric currents have magnetic properties and that magnets will move in response to a changing electric field, properties that led to the inventions of the electric motor and the dynamo; the electric motor uses electrical energy to produce mechanical motion, and the dynamo uses mechanical energy to generate electricity. An electric motor could run on a battery, but the batteries available early in the century did not store much energy, so there weren't a lot of practical applications at first. New inventions continued to improve battery technology, finally producing batteries that were relatively efficient and capable of being mounted on a vehicle. Dynamos had to be driven by something, so the steam engine was an obvious candidate. Inventors like Thomas Edison recognized that copper wires could carry electricity from a steam-powered dynamo to a distant electric motor that could do some useful work. Edison also recognized that developing a practical light bulb was just one step in the process of lighting buildings. To do this required a practical electrical generator, a practical means of getting electricity to the buildings, a safe way to wire the buildings, and a practical means of hooking up the wires to the light bulbs. This monumental task involved technological advances in several different areas. Hundreds of inventors tested out thousands of potential solutions before it could be done safely and economically.

In 1882 Edison put it all together and built the Pearl Street power station in New York City, initially serving eighty-five customers.[5] Within two years it was serving five hundred customers with ten thousand light bulbs. He used the latest steam-engine designs to power the dynamos. A drawback of his design was the limited range afforded by his choice of direct current (DC). Replicating his design in cities would have required a huge number of power stations and would not have been practical for household use in rural areas. George Westinghouse bought the patents for alternating current (AC) generation from Nicola Tesla, and in 1893 the Chicago World's Fair was lit up by his system. With AC it proved to be possible to transform the electricity to very high voltages, whereby the current could be transmitted long distances with much less loss of energy than with DC. Today AC is the primary technique for the transmission of electricity, primarily using steam turbines driven by burning coal, natural gas, and nuclear fuel.

Once electricity became available, a source of power was available to each business and home, power that could be used for other things besides light bulbs. We will see how this stimulated many inventions in the next century that had major lifestyle implications.

In 1833 Carl Gauss and Wilhelm Weber used two wires to transmit hand-generated electrical pulses from a coil and magnet for a distance of about one kilometer (six-tenths of a mile) to a receiver, the first electrical telegraph. This demonstrated that information could be transmitted quickly over long distances. In 1837 patents were taken out independently in the United States by Samuel Morse and in Britain by William Cooke and Charles Wheatstone. In the next few years, telegraph lines were installed all over the United States and Europe. In 1866 an oceanic cable was laid between the United States and Britain. Messages that previously would have taken a week to deliver suddenly took only a few minutes, a capability that would have far-reaching consequences. Once the lines were laid, inventors like Thomas Edison worked hard to find ways of speeding up the information and improving the reliability and range of the telegraph signals. Later the telephone was invented and then progressively improved until it became a practical device by the end of the century.

By the 1800s new ideas were rapidly transmitted around the world, abetted by communication between mother countries and their colonies. While the initial inventions originated in Europe, rulers and local authorities everywhere recognized their utility. By the end of the century, telegraph lines were installed in South America, Russia, the Middle East, India, Japan, China, Indonesia, Australia, and New Zealand. Sweden was the early leader in the installation of telephone services, along with New Zealand, Switzerland, and Norway.[6]

As with most new capabilities, the telegraph and telephone were initially used by wealthy people as novelties and by businesses that could profitably use the increased speed of transactions. For reasons that will be discussed later, the early uses were widespread enough to encourage further innovations by competing parties, the result being that the cost of sending information decreased over time. The reduced cost, in turn, enabled a larger segment of the population to utilize the service for quickly communicating information about personal events.

One of the most important innovations of the century did not involve inventors or scientists, but rather businessmen and politicians: it was the corporation, or (as it is known in Britain) the public company. It first came into existence in Britain with the passage of the Limited Liability Act of 1855, which was soon followed by the Joint Stock Companies Act of 1856 and the Companies Act of 1862.[7] With these acts, businesses could be organized as corporations with stockholders whose personal assets were protected if the company went under. This was extremely important to attracting a large number of potential stockholders who, while they had some savings, would not be willing to risk their homes and properties. With limited liability, the creditors of the corporation could go after the assets of the company—which included the stockholders' investments—in case of bankruptcy, but they could not go after the stockholders' assets (i.e., their homes and their bank accounts). Prior to this act, owners and shareholders were "betting the farm" if they invested in a company that seemed to have a good chance of being profitable. This meant that only rich people with enough assets to cover losses of a company were able to participate, and it severely limited the number of people who could benefit from a successful enterprise. By limiting the risk to investors, a corporation could attract much more investment and become larger in response to its success in the marketplace. Less wealthy people could become richer if they invested in successful companies, and they would know that, at worst, they would only lose their investments if things went sour. As John Kay pointed out,[8] the corporation enabled stockholders to invest in large organizations run by salaried managers. It made it possible for wealthy individuals to provide capital to businesses run by entrepreneurs and professional managers without having to be involved with day-to-day operations. It also opened the door to less affluent stockholders to invest money in the economy.

Within two decades similar laws were passed in the United States, Europe, and many other countries, and these public companies proliferated. A recent column in *The Economist*[9] calls the public company a type of organization that has spread prosperity and created wealth for the last 150 years. It notes that public companies have been crucial to the creation of jobs and to innovation, that they bring with them a welcome and beneficial transparency, and that they give ordinary people an opportunity to gain from their investments. Over the fifty-year period from 1820 to 1870, worldwide GDP rose by about 50 percent, from GK$700 billion to GK$1,100 billion, but in the thirty years from 1870 to 1900, it almost doubled to GK$2,000 billion. How much of this spectacular growth was due to the creation of the public company is difficult to assess, but there is no doubt it played a major role.

The Twentieth Century

It would take books to even summarize the inventions of the twentieth century, so I will confine the discussion here to a few of the more obvious ones. By this century the benefits of innovation were taken for granted, and people began to believe that innovation would continue forever; it certainly shows little sign of slowing down yet. During and immediately prior to World War II, governments poured huge amounts of money into research on technologies that were connected with the war effort: jet aircraft, rocketry, radar and navigation techniques, information theory and encryption, computing machines, and the atomic bomb. Motivated by patriotic fervor, innovators achieved in a few short years technological advances that were nothing short of phenomenal—achievements that formed the foundation of many civil and commercial advances of the last half of the century: jet airplanes, positioning and navigation systems that included the satellite-based Global Positioning System (GPS), radar techniques that led to air-traffic control systems, communication satellites, microwave transmission of data from towers and satellites, and the transistor, the integrated circuit, and the computer. These were all supported by important scientific developments in information theory, cybernetics, fluid mechanics, and applied mathematics, to name but a few.

In the world of entertainment, radio, records, and movies came on the scene early in the century, and people everywhere immediately used them. Radio stations proliferated, broadcasting news, comedies, westerns, and music. Movie theaters were constructed in cities and towns worldwide, and they became popular with all ages. People bought record players and were then able to listen to the best musicians from all over the world. These were all subsequently improved in quality, cost, and availability. They were succeeded by television, tapes, CDs, DVDs, portable players using digital storage, and the internet. Today movies and television shows can be streamed directly onto a computer or television using high-speed data links.

By the twentieth century it was widely recognized that inventions were engines for economic growth and new possibilities of all sorts, so governments and companies looked for ways to support innovative efforts. Companies formed internal research and development (R&D) departments to create new products and improve them. Research centers with broader goals were established with government and private funds. The object was to systematize invention by providing appropriate environments. The notions of "basic" and "applied" research came into vogue, echoing the white-collar/blue-collar and science/technology distinctions. Companies were looking for near-term benefits, so naturally they mostly supported activity aimed at developing commercial products; the time horizons of basic research and even most applied research were too far out. Governments could and did take a longer view and funded efforts to develop technologies that had potential payoffs in future decades. Research centers were set up around universities and competed for government and private funding. University departments of science and engineering conducted basic research activities without having a specific final product or capability in mind; instead, projects were oriented toward an improved understanding of nature. Individual inventors working with individual entrepreneurs became increasingly rare.

While the library and museum at Alexandria was one of the first research institutes that brought innovative people together in one location, Thomas Edison's Menlo Park laboratory in New Jersey was one of the first research institutes devoted to the invention of devices and concepts useful to the public. It was specifically designed to elaborate on his ideas, and he closely controlled it. World War I spurred practical improvements in airplanes, vehicles, and ships oriented toward weapons and logistics. During World War II, research was carried to a new level: while companies were funded to improve and adapt existing war-related technologies, government laboratories like the Massachusetts Institute of Technology Radiation Laboratory (MIT RadLab) were set up to develop new technologies like radar, navigation, rockets, and atomic weapons.

They employed thousands of physicists and engineers who accomplished in an unbelievably short time a number of impressive breakthroughs that affected the outcome of the war, including radars that were used effectively on ships and land.

By the time the war ended, the physics involved with these technologies were much better understood by a large number of very smart people, and it was clear that the possible peacetime applications were enormous. The stage was set for the explosion of technology that subsequently took place in data and voice communications, navigation, jet aircraft, rockets, lasers, semiconductors, integrated circuits, computers, appliances, medical electronics, and exotic materials. While the US Department of Defense continued its support of military technology development, many of these areas were accelerated by the NASA moon-landing adventure and its requirements for highly reliable, vibration-resistant, lightweight equipment. Research centers proliferated to take advantage of the new possibilities and the newly available talent. As these capabilities were exploited to provide useful products to businesses and consumers, companies also established research centers and supported R&D efforts that showed potential for future products. As a result there were many sources of support for innovative activity.

This innovative activity was spread across the spectrum of research—from theoretical to practical, from basic to applied, from conceptual to commercial. A great deal of thought was put into how best to harness the available brainpower and enthusiasm, resulting in different approaches for different kinds of research. Product-oriented research tended to focus on using new materials and processing techniques to improve speed, increase capability, use less energy, reduce weight, reduce production costs, and improve reliability. Applied research was more oriented toward understanding the physical processes of useful technologies and producing potentially useful materials and techniques. Basic research focused on improving our understanding of the physical world, with less emphasis on useful technologies.

In establishing research centers to conduct basic and applied research, some forward-looking managers recognized that Edison's Menlo Park laboratory was not a good model to follow. There, the lone genius decided what topics would be explored and who would explore them, and he had very specific ideas about the applications of the inventions that were produced. Highly innovative people want to learn from geniuses, but they are not content to follow direction. As a consequence, they need a more open and less structured environment to do their best work. My colleague Brad Parkinson, who is called the father of the GPS[10] because of his pioneering technical and managerial contributions to its development and deployment, once described an ideal approach to a successful research center in a talk at the US Institute of Navigation.[11] By applying the principles of what he termed "Cannon's law" (in honor of his mentor Bob Cannon at Stanford University), the innovations that would be produced would be both different from and superior to the initial expectations. A critical step in making this happen is to hire the best people possible in the field of endeavor. The next step is to provide an initial focus for the effort. The third step is to get out of the way. The statement of the law is "one thing leads to another," which at first sounds simplistic, but it contains the recognition that significant innovations do not develop in a straight line. Often what results is quite different from the initial expectation; the goals will almost surely change during the course of the work. This approach fully meets the essential conditions for innovation that are developed later in this book.

Basic, applied, and product-oriented research areas frequently overlap, and a number of important developments occurred at these intersections. Claude Shannon at Bell Labs, for example, developed a basic relationship in information theory between the amount of data that could be transmitted in a given time over a channel and the ratio of the strength of the electrical signal carrying the data to the amount of inevitable noise in the channel, independent of the technique used to encode the data.[12] This theoretically derived limit had two significant consequences for the engineers designing data channels and encoding techniques: (1) they could evaluate alternative designs against the theoretical limit, and (2) they could avoid wasting their time trying to develop designs to exceed it. Data channels are used by all communications equipment. Another example that shows how practical results can feed into basic research was another effort at Bell Labs, an antenna and radio receiver development.[13] Arno Penzias and Robert Wilson were developing an antenna and receiver to detect very weak radio signals bounced off metalized echo balloon satellites. After doing everything they could think of to eliminate noise sources from the receiver and the antenna, they still found there was a significant level of noise they couldn't account for, and it stayed at the same level no matter where the antenna was pointed. Later it was determined that this was exactly the noise level predicted by the Big Bang theory of cosmology. This observation, repeated many times, is one of the central pieces of experimental evidence supporting the theory. There are many other examples of such cross-fertilization between different kinds of research.

The growth in research centers and academic departments was matched by an explosion in scientific and engineering journals, all of which became increasingly focused and specialized. With this increased specialization and the explosion of technical and scientific knowledge, it became impossible for researchers to keep track of progress in other fields, and they had little contact with researchers in those fields. There have recently been a number of attempts to encourage cross-fertilization between disparate fields in order to reduce this isolation of thought. An early example was Building 20 at MIT,[14] where it happened serendipitously. This large three-story building, left over from the RadLab days, became variously occupied for several years by researchers in the disparate fields of linguistics, acoustics, electronic circuit design, high-speed photography, anthropology, science-teaching techniques, and biology. The researchers there mingled with each other; they argued, discussed, and shared their ideas. It was a stimulating environment that was referred to as "the magic incubator" by people at the university.[15] Today there are many centers that encourage this type of interdisciplinary contact.

During the twentieth century there were a number of attempts to use groups to develop innovative ideas. Groups comprised of individuals who have complementary skills have been successful in decision-making because the collective experience of the groups enabled them to weed out bad ideas. Teams have also been effective in product development for many of the same reasons. Many in the United States were surprised by the success of product development in Japan, where teams were used extensively. As noted above, contact with other people, even in other fields, can stimulate new ideas. However, it is difficult within a group to work out the details essential to significant innovations, and such efforts have not been very successful.[15] Individuals are at the heart of all innovative activity, and they need time to explore ideas at a deep level and to conduct the necessary research in solitude.

Summary of the Last Millennium

Before addressing how innovation was nurtured by and, in turn, impacted political and social developments, let's sum up the developments of the last thousand years. During this period, increased trade resulted in a growth of wealth in several areas of the world. In China and India, this was matched by a proportional growth of population, so the GDP per capita did not change. Population growth was spread across the classes, and there was little economic upward mobility. However, in the Italian city-states, the trade increased faster than the population growth, creating opportunities for lower-class citizens and encouraging more widespread literacy and education; the GDP growth more than doubled between 1000 and 1500 to about GK$1100 per capita, after which it leveled off. A similar phenomenon happened in northern Europe soon afterward. By 1600 the Dutch overtook and surpassed the Italians in GDP per capita until it reached more than three times the world average (GK$2,000) by 1700. There were unique changes that took place in Britain that created a favorable environment for inventors, who—working with entrepreneurs and businessmen—produced innovations that not only were self-sustaining but also encouraged new innovations, what we in engineering call a "positive feedback loop." Each new invention spawned new possibilities, and there were enough literate and educated inventors working with competing businessmen that the heavy hand of the monarchs and monopolists couldn't control the situation. Once the successes of the new technologies in Britain began to be evident, the lesson spread to other parts of the world. The Low Countries already had a middle class with capitalist values, and they immediately began to use and adapt the emerging technologies. It reached the point where even the people at the top of the economic ladder all over Europe began to recognize that they too were benefiting, and new jobs became increasingly available—jobs that eventually created a middle class and a wealth that was more widely distributed; this led to more stable societies and more pleasant environments. There was a widespread recognition that innovation resulted not only in interesting toys and gadgets but also in successful economies. The countries that achieved democracies with institutions that protected entrepreneurial activity prospered, while others did not. Today it is widely recognized that among the different forms of government, democracies have out-innovated and out-performed the others. As messy as they are, they have proved to be stable while providing their people the widest array of freedoms.

The next chapter will explore how a unique set of mutually reinforcing conditions collectively caused this explosive growth to happen.

Part II: How, Where, and Why Innovation Occurs

7. Invention Spurs Invention

Technology begets technology—a perfect storm of changing conditions; the empowerment of individuals; competition within societies; the appearance of the modern market; size matters; innovations that promote democratization

In his important book *Guns, Germs, and Steel*, Jared Diamond claims that under the right conditions, technology begets more technology, a process he calls "autocatalysis."[1] He cites as a prime counterexample the Cretan Phaistos Disk,[2] a clay disk dating back to 1700 BC that had punched symbols on both sides. The several dozen symbols are frequently repeated, indicating that a stamp was used,; this suggests that many such disks were crafted. The meaning of the story told on the disk has not yet been determined. Diamond raises the question as to why this seemingly important innovation was not replicated, as it has many of the features of the printing press. He concludes that it was not replicated because it was invented in isolation, whereas the printing press was able to make use of several innovations not available to the ancient inventors from Crete: paper, movable type, metallurgy, presses, inks, and alphabetic symbols.[3] The printing press was a synthesis of these already existing and widely used techniques. Once the first prototype was built, the materials and techniques were there to replicate and improve the initial design. The Phaistos Disk was not invented in that kind of environment.

What Diamond calls "autocatalysis," David Landes calls "the invention of invention," that is, the recognition that the process of invention is itself subject to innovation.[4] While Diamond cites the reasons given above for the success of the printing press, Landes primarily credits the existence of a market in Europe. There were other elements involved in the perfect storm of developments that gained momentum after 1700 and which together formed a serendipitous environment that enabled innovation to take off. Collectively, they were the following:

- The increasing empowerment of a growing portion of the population, made possible by the progressive loss of power of the monarchies and the increasing power of parliaments.
- The rapidly expanding knowledge in technology, availability of resources, and science, fed by increasing literacy and education of the public.

- The increasing perception of inventions as being desirable for the economy of a society: by monarchs, landowners, merchants, and most of the rest of the populace. What also changed was the recognition that it is in the public interest that new ideas be made available, rather than kept secret.
- The emergence of the modern market and competition within societies as well as between them, which resulted in cheaper goods and increased consumption.
- The democratizing and empowering effect of certain innovations, forming a positive feedback loop, which led to further empowerment and thus to further innovations.

Each of these elements increased significantly beginning in the 1800s, and they continue to grow today.

Empowerment

It was noted in the last chapter that in Europe the power of the monarchs to control and intimidate the populace was considerably weakened by the loss of religious mystique, by the power of the landowners, town fathers, and the rising merchant class, and by the importance of craftsmen and artisans to the economy. As more people claimed the power to be heard, the notion of universal rights by birth was born, along with the idea that governments derive power from the governed; such concepts would have been unthinkable in the old autocracies, where people were considered to be merely subjects of the ruler. Elected parliaments such as Britain's House of Lords and House of Commons, which were initially established to assist the monarchs and do their bidding, gradually acquired more and more power until by around 1900 the monarchs were reduced to figureheads and symbols of national pride.

Increased empowerment also came from expanded concepts of property that took root initially in Britain. Patents could be owned by anyone, and patents were a form of property in the sense that the ideas expressed in them belonged, at least for a period of time, to the inventor. As more people recognized the possibilities, dreamers came out of the woodwork. Rosen[5] said it quite eloquently: "Once a generation of artisans discovered they could prosper from owning, even temporarily, the fruits of their mental labor, they began investing that labor where they saw the largest potential return. Most failed, of course, but that didn't stop a trickle of inventors from becoming a flood." Land and commercial buildings remained mostly in the hands of the aristocracy, of course, but in the colonies land was made available to the adventurous, and people responded. Once they could keep the fruits of their labors, they worked hard, and the economies grew. In the colonies of America, Canada, and Australia, annual GDP per capita rose from the subsistence level of GK$400 before 1700 to about GK$1200 by 1820, matching western Europe.

Monarchs, of course, had little direct contact with lower-class people (except for servants), seeing them primarily as a source of labor for farms, mines, ships, and the military. Similarly, landowners saw them as ignorant illiterates who were there to work the farms, forests, and mines. The merchants saw them somewhat differently because they needed specially trained people who could operate looms and other machines, handle goods being exported and imported, keep books, and help sell products. However, under mercantilism the workers continued to be treated with scorn by the upper classes and were kept illiterate and poor. The primary mercantilist goal was to increase the quantity of manufacturing goods for export, and little else mattered. In response to this, there was widespread worker resistance, and the lower classes gained more economic and political power; some colonies rebelled and achieved their independence. Literacy increased, and people got to read firsthand documents—like the American Declaration of Independence—that proclaimed their natural rights. Over time, the base of the electorate of legislative bodies became broader; later it included men who did not own property, and finally women. Once people felt that they were important, they lost much of the fear that the old autocrats had worked so hard to instill. As more people lost their fear of the authorities, the pool of potential innovators grew. Chapter 10 discusses the effects of fear on innovative activity and the empowerment people feel when this fear is removed.

The Expansion of Technical Knowledge and Literacy

Once the printing press was invented and it became possible to print books, magazines, newspapers, and pamphlets cheaply, everyone wanted to read, and the desire for widespread literacy eventually won out against those in the ruling classes who wanted to suppress this dangerous development. Blue-collar, as well as white-collar, knowledge could be recorded and distributed. Here again, the forces that wanted to make the knowledge available won out against those who wanted to control it for their own purposes. Patent laws played a key role because, while they provided for a short-term monopoly on inventions, the technological knowledge contained in the patents became public. That knowledge was then used to improve upon the ideas and inventions in the patents.

Scientific knowledge was originally a white-collar affair. Most scientists, or natural philosophers, were well-to-do, titled people who were intelligent and curious and had time to consider scientific questions of the times and perform experiments. They were not supported directly by government grants, although many of them held academic positions at universities. It was in London where the first journal of scientific articles, the *Philosophical Transactions of the Royal Society*, was published in 1665. To be a fellow of the Royal Society, one had to be a "gentleman, free and unconfin'd."[6] To its credit, the Royal Society considered the extraordinary commoner Robert Hooke to be so talented and indispensable that it gave him a salaried position in 1665, although he was excluded from being a fellow. The fact that he was "confin'd" by his need to make a living was overlooked in his case.

Scientific and technological societies soon sprang up all over the world. They published articles in magazines, periodicals, and books. By the end of the 1800s, there were publications in just about every human endeavor imaginable, all adding useful hypotheses, theories, and experimental results to the collective body of knowledge in each subject. These continue to expand and proliferate to this day.

With computers and the internet today, it is possible to quickly find information on any subject and to store it for later reading and analysis. The availability of this information is ubiquitous, a condition that would have astounded people fifty years ago. As someone who began conducting research decades ago, I am still in awe of the new capabilities—questions that would have taken weeks to explore can now be answered in minutes, or less.

Until the 1800s, literacy was confined mostly to the upper classes; even when educational services were, in principle, available to others, few could afford to take advantage of it. There was little incentive for peasants working on farms to learn to read, and most people lived on farms. In towns, things were different; with increased trade, the merchants needed workers with at least rudimentary reading and arithmetic skills, so there were some incentives. Craftsmen found that they could benefit from instruction manuals, and Protestants in particular wanted to be able to read the Bible. Soon there was widespread interest in the concept of universal education; John Adams, Ben Franklin, and Thomas Jefferson were among the Founding Fathers to advocate it in the United States, and proponents were everywhere. They succeeded in getting the concept accepted to the point where very few people openly advocated keeping anyone illiterate. While it's difficult today to believe that anyone would do so, it was the norm in the past.

As economic life became more complex, the rewards for training and advanced education became apparent, and parents increasingly encouraged their children to pursue education beyond high school. Colleges and universities offered training in professions, including engineering, as well as in the liberal arts. The building and craft trades supported training programs for young people.

Chapters 11 and 12 address the impact of access to information and education on innovative activity in more detail.

The Changing Perception of Invention

Among literate people, who before 1800 were a small minority of the population anywhere, the virtues of scientific endeavor were widely recognized and encouraged. This was not true of inventions, which were often referred to as "gadgets" and "toys."[7] Most of Leonardo da Vinci's inventions were not practical, often because the materials available to him weren't adequate, so his accomplishments didn't change that attitude. The autocracies of the East treated clocks as curiosities rather than practical instruments. Most aristocrats considered inventive activities to be rather undignified, the work of "low mechanicks."[8] This attitude gradually changed as scientists found useful applications of their work. Galileo was one of the earlier ones, followed by Robert Hooke, Robert Boyle, Benjamin Franklin, and many others.

The monarchs were interested primarily in inventions that provided a source of revenue for their military adventures and other royal expenses. They were not likely to be a source of capital for the development of an invention unless success was virtually assured or if the potential payoff was large, as in the case of Christopher Columbus obtaining support from Ferdinand and Isabella of Spain. The landed aristocrats were fairly conservative as well; they were mostly concerned with products directly associated with their lands.

In Europe, another class of people emerged—namely, the burghers and officials who governed the growing towns and cities. These families ran businesses and played the feudal lords off against the monarchs to gain power and get themselves represented in governments, one example being the House of Commons in Britain. Some of them came from landed families (second and third sons who could not inherit land because of primogeniture laws). They were often business-oriented, and they looked for ways of increasing the family fortunes. As an example, a wealthy burgher in Mainz named Johann Fust loaned Gutenberg money to set up a printing press and print Bibles and other documents. In a story that was to be replicated many, many times in history, Fust sued Gutenberg and took over control of the profitable enterprise, leaving Gutenberg bankrupt.

The emerging merchant class was even more enterprising. With their access to markets outside the kingdoms, members of the merchant class recognized that they could become wealthier if they could produce and sell more goods. They were motivated to find more efficient ways to produce goods in larger quantities and more cheaply. Since they were dealing with customers in many countries, they were competing with other merchants who just might find ways to offer the same goods at a lower price. If this happened, they could go bankrupt. By the same token, if they could find better production methods, they might be able to put their competitors out of business. In such an environment, inventors who could develop machines or techniques to help the merchants succeed would find a receptive audience. Plus, this audience was not just one agent of the king; there were many merchants, and if one was not receptive, the next one might be.

The idea that inventions could place a merchant in a better competitive position was a powerful incentive for innovation. This concept powered the Industrial Revolution once the brakes imposed by the monopolistic control of the monarchs and landowners were sufficiently weakened and as merchants, town officials, and businessmen became more politically powerful.

The merchants, businessmen, burghers, landowners, and monarchs came to see their interests as allied rather than as a zero-sum game where for every winner, there must be a loser. This alignment was given a focus with the concept of the "nation" and its associated "national interest." The idea was that all classes could benefit by acting to promote the national good, and thus mercantilism was born. By increasing the manufacturing production of a nation, and by exporting the products, the nation could become wealthier and more powerful. As a corollary, this idea implied that importing products would weaken the nation. So mercantilism had features that encouraged invention, but it also had anti-market features held over from the monopolistic mentality.

The encouragement of manufacturing was a boon to inventive activity that increased production. It also aligned the perception among the now-expanded classes in power that invention was good. It additionally had the effect of promoting the notion that new ideas and inventions should be made public, at least within the nation, so as to stimulate better inventions and improvements on old ones. It incidentally resulted in an increased social mobility, as wealthy merchants bought baronetcies or married nobility, and landowners and burghers went into business.

The one class of people that didn't share this enthusiasm was the lower class. Mercantilism needed a large number of people to work the factories and mines, but the workers were seen as a form of natural resources of the kind that could be allowed to increase in size as needed. They were expected to procreate so that their children could take over the routine work; they were not seen as sources of future innovation. Many jobs were eliminated and replaced by machines, and the plight of most workers actually got worse, as they were expected to work longer hours than before and to perform more repetitive work—and at reduced wages. The Luddites of the early nineteenth century were among the better-known protesters of this situation. Their anger was particularly aimed at the use of automated looms that replaced skilled textile workers with cheap labor. They destroyed many of these looms, as well as cotton and wool mills. Eventually, however, the benefits of automation brought measurable improvements to the standard of living of most people, and the idea of productivity gains being ultimately good for everyone became accepted by most of the population.

The Emergence of the Modern Market

It is worthwhile exploring what "market" means because it is crucial to understanding why the self-perpetuating process of "invention spurring invention" occurred when and where it did.

A "market" can be a gathering place where people go to buy and sell (or barter) different kinds of goods and food. This type of market has existed for millennia and in all kinds of societies. It played a key role in the development of towns in medieval Europe, as merchants brought their wares from abroad to different sites. However, when we speak of a "free-market system," we are not referring to this kind of market.

"Market" can also mean the potential number of sales for a given product—we say there is a "worldwide market" for computers, for example. The size and accessibility of the market for a product is important for inventors. If an inventor lives in a relatively isolated small town, then the market for a useful gadget will be quite small. In such circumstances, an inventor may be motivated more by community status than by money. In the Shaker community around 1800, there were a number of inventions that helped the community raise food and build houses, among them an improved circular saw. The inventors here were inspired by community and religious motives, rather than by the potential of becoming rich. In a city or large community, there will be more opportunities to make a profit from a useful invention. Merchants who traveled through towns and villages acted as middlemen, so inventive people could learn of potential markets outside their own areas.

The potential for profit from an innovation depends on the potential market size in absolute terms—the bigger the potential market, the more the incentive to spend the time and effort to develop a product. Bigger markets justify a larger investment of money up front. A large potential market is also a strong motivation to produce the product rapidly in higher quantities, which in turn is a strong incentive for inventing new machines and techniques. Bigger markets make it more difficult for a single merchant or producer to dominate the sales of any product. Whenever there are several producers, they will compete unless there are laws or controlling forces that prevent it. Once there is adequate competition, no producer can afford to charge significantly more than the others, and there is an additional incentive to find ways to reduce the costs of production (i.e., to increase productivity). That way, more of the money from sales goes into the producers' pockets. So in such circumstances, people are motivated to invent products that other people will want; the manufacture of the product will motivate inventions and improvements to increase the volume of and reduce the costs of production. When there is a growing market for a variety of products, there is an additional motivation for innovations that reduce the costs of shipping and speed up the shipping time. Bigger markets also generate more variety, because markets tend to segment. For example, textiles from a given factory will be produced initially in a few patterns and colors. The customers, while they may parade about showing off a certain popular style, will soon express the desire for more variety—more colors, wider variations in the patterns, finer textures, coarser textures, different materials, etc. In responding to the changes in demand, the manufacturers will need modified looms and other machines, all of which call for innovation.

In the autocracies this kind of market for most new products or ideas was limited to one set of customers: the ruler and his agents. Consider the options for an ironworker who imagined a new tool for shaping chariot wheels, and suppose he somehow developed a prototype on his own. Who would be interested in using his invention? What reward could he realistically expect to get from it? Most likely, the interested party would be an army officer who would see the benefit of making chariot wheels more quickly and reliably. If the inventor was lucky, the officer might give him a special position where he could supervise the fabrication of chariot wheels for the army. On the other hand, the officer might claim credit for the idea himself and leave the inventor in the lurch. In either case, the army officers would want to keep the tool a secret because it would give them an advantage over their enemies. The inventor would not be allowed to sell his idea in other countries or to makers of nonmilitary chariots.

One type of market for some goods has existed since ancient times—namely, the inter-kingdom markets associated with trade. Rulers usually found it more convenient to trade goods between them than to try to use military means to capture the source of these goods, although when they were considered valuable enough, conquering and taking over those resources was a time-honored means of bypassing the middlemen. It is said that Solomon's hundreds of wives were the daughters of neighboring kings; this arrangement enabled his merchants to travel with a minimum of tolls and taxation along the trade routes. While this story may be apocryphal, there certainly were many merchants in ancient times that traded goods between the kingdoms. While this trade was mutually beneficial to the respective societies, most of the benefits went to the upper classes. They controlled the production of the goods that were traded out of the kingdoms. An individual inventor who had an idea for improving the quality of those goods might expect some reward from the family who controlled their production, but little more.

While most merchants were agents of the rulers, some achieved considerable independence and were able to establish fortunes of their own, as happened in ancient Greece. There and later in medieval Europe, some merchants became rich and coordinated enough to form a separate class with some clout. Elsewhere they continued to be agents of the emperors and local kings.

Today we use the term "free market" to describe the exchange of goods and money between countries free from excessive tariffs and taxes. In a free market, prices for goods are not fixed or prescribed by law, but rather they fluctuate to respond to the variations in supply and demand. If a particular product suddenly becomes very popular (i.e., "in demand"), the supply may be, temporarily at least, inadequate to fill the demand, in which case the price will rise. With a rise in price, some people will decide the product is not worth buying, and the price will adjust until supply and demand are balanced. Conversely, if people tire of a particular product, the loss in demand will result in a price drop, which reduces the incentive to produce it. If the supply of a product is limited by the availability of natural resources, the price will rise until the demand drops to the level at which purchases can be sustained. Similarly, as new sources of natural resources are found, prices will drop. The idealized situation described here is called "perfect competition" and is a reasonably close description of how many products are bought and sold, especially commodities involving large markets.

Today there are significant sectors of every economy that don't follow these idealized laws. Some countries and companies engage in "dumping," in which the price of a product requiring a large initial investment is artificially held well below its free-market price in order to drive one or more competitors out of business. Once out of business, the victim competitors find it difficult or impossible to recover, during which time the prices are raised back to or above the free-market price by the dumper. Governments sometimes intervene by placing taxes on certain goods in order to control the use of products such as tobacco, or they may place tariffs on imported goods to protect local industries. Most societies today have economies that are a mix of free markets and controlled ones.

Things were different up to about two hundred years ago. *There were few free markets of any size prior to about 1800; much of the commerce internal to societies was dominated by monopolies, protected by law.* Competition was considered destructive, something to be avoided. The rulers were motivated to subdivide the economy into enough separate categories to keep peace among loyal families and to raise revenue, and they used their power to reinforce loyalty and punish dissenters.

Mercantilism continued some of the features of monarch-granted monopolies, but the sheer size and variety of the growing trade made it more difficult for the monarchic bureaucracies to keep up with the changes and to maintain control through granted privileges. These monopolies also conflicted with the interests of the merchants and businessmen, who had become more politically powerful. By the end of the eighteenth century, the writings of Adam Smith (*The Wealth of Nations*) and others who opposed mercantilist theories had some effect. Whatever the major causes, more and more goods came to be traded outside the monopolies, and competition between merchants within the same countries increased. This was a key development.

By the end of the nineteenth century, David Ricardo's concept of "comparative advantage" was understood and accepted by most European nations and their offshoots in the newly independent colonies. This theory posits that international trade between two parties is not a zero-sum game, but rather can be a transactional situation in which both sides benefit. Trade barriers, cross-border taxes, and tariffs were significantly reduced in Europe as the benefits of doing so became more widely recognized.

The Impact of Size on Innovation

It was noted above that the absolute size of the market influences the opportunities for product innovation. It is worthwhile expanding on this concept because it has important consequences.

In tribal societies any new capability would find use primarily within the tribes, which were quite small. The kinds of innovation were thus incremental ones that made life a little easier (e.g., better tools and weapons, better food containers, and easier means to light fires). Tribes that traded with each other learned what objects the others liked, and that stimulated improvements. When civilizations came on the scene, the leaders took on larger projects (e.g., large buildings and tombs, bridges, and even whole cities). These vastly expanded the opportunities for creative and innovative people, but they were still limited by the autocratic desires. In this case the greater size of the population had the effect of increasing the variety of opportunities for innovation but didn't create much in the way of markets, as we think of them today, because the demand for products was largely dominated by the autocrats and the elite families. The size of the land area controlled by the autocrats brought into play natural resources. Different autocrats controlled different kinds of foods, woods with different properties, mines with different minerals, and quarries with different materials. These not only were useful for projects within societies, but also supported a brisk trade between them. Variety begets variety; as people recognize the benefits of different types of any product, it is only natural to ask whether there are other possibilities. If cloth comes in two different colors or weaves, might there be three or four or more? Variety creates opportunities for innovation.

Fast-forward to the Middle Ages, and the world population had grown significantly. There were a number of relatively large cities throughout the world, and towns were forming all over Europe. The elite of the

Italian city-states competed with each other to wear the best clothes, build the most impressive estates, eat the best food, and sponsor the best music and art. This situation created a demand for variety in all manner of goods and services. The number of European merchants had proliferated, making it possible to obtain products from all over the world. In turn, they needed products to trade, so piecework fabrication of tradable products grew. By this time, merchants had access to much larger markets, so innovations that could make possible larger volumes of products quickly were encouraged. Eventually this led to centralized factories powered by wind, water, and steam. As the markets between kingdoms grew, so did markets within the kingdoms.

When the market for a product is significantly larger—where instead of selling a few dozen, one could sell thousands—it opens the door to major innovations that require considerable investment up front. As a simple example, if a dressmaker can sell only a few dresses per month in a small town, sewing them all by hand makes sense. If the demand is for a hundred each month in a city, it's worthwhile to purchase a sewing machine. If the market is for thousands for sale to other countries, it's worthwhile spending the money up front to purchase automated equipment. Size makes a difference; larger potential markets are a spur to inventive activity.

Democratizing Innovations

Democracy is not a governmental philosophy handed down by beneficent rulers. Rulers have historically considered their legitimacy to come from a higher authority, while democratic governments derive their power from the people. Such words would have been anathema to the autocrats, who typically claimed their higher authority to be the gods. When the power of the Catholic Church broke the association of the European rulers with heaven, the monarchs that followed had to rely mostly on force and control, although they attempted to enforce the "divine right of kings." They tried to control the information channels and to limit the ability of people to organize opposition movements or to move out of their area of control. This was best accomplished by keeping people poor, restricting their physical and economic mobility, restricting their ability to organize, and enforcing their class identity (i.e., "keeping them in their place"). The will of the people had little place in this type of societal structure, and democracy and the power to govern had to be wrested from the reluctant hands of the ruling classes or removed by stealth. Democratizing innovations are those that provide means to empower the general populace.

There are four types of innovation that do this, and each one has elements of stealth because the reasons why they are democratizing were not evident to the ruling classes when they were first developed. By the time their effects were apparent, the principles and benefits of democratic government were widely accepted and were no longer easily controllable or reversible. The four classes of democratizing innovations are:

- Transportation
- Distribution of information
- Communication
- Entertainment

What these kinds of innovations do is to greatly increase the number of options available to people. They expand people's choices of where they can live, where they can work, for whom they can work, to whom they can talk and listen, and how and where they can spend their spare time. Modes of transportation enable people to visit other people and places, conduct business, and relocate. Widespread distribution channels give people the information they need to make informed choices in every aspect of life (i.e., politics, education, religion, employment, and social and leisure time). Rapid communications enable people to pack many more experiences into their lifetimes. Different kinds of entertainment add to the quality of life. All of these innovations empower people and give them a sense of responsibility and self-importance. They enable people to organize and resist oppressive governments and institutions. They give people the message that they are important and that their efforts can make a difference in the world. Innovators thrive in such an environment.

These types of innovations are stealthy, in that their democratizing potential is not immediately apparent. Except for entertainment, where in a relatively short time the ancient Greeks created the concepts of amphitheaters and stadiums for all posterity, these areas of innovation began with inventions that primarily benefited the ruling classes, with little initial trickle-down to the lower classes. New materials and crafting techniques supported the building of great tombs, palaces, and cities, but most people lived in homes made of the cheapest materials available and with limited furniture and decorations. They didn't ride in carriages, and they wore modest clothing. Stated in modern terms, there were very few items that had a mass market.

Transportation innovations began with improvements in ships, animal-drawn carts, wagons, and carriages, which were widely used into the nineteenth century. Locomotives were used initially for transporting materials, but they soon carried passengers as well, and in a few decades they were available to travelers across the class spectrum. Electric trolleys were used in cities for mass transportation. In a few years the automobile was invented and frequently improved and changed, and it became widely used. Automobiles were followed by buses, trucks, and farm and construction vehicles. The airplane now sees wide use, especially by middle- and upper-class people. These methods of transportation now provide a wide variety of options of places to work, live, and visit. The different modes of transportation made it more difficult for the ruling classes to limit the mobility of the ordinary people.

In ancient times representatives of the rulers made public announcements centrally. This mode of information distribution was common, even after the invention of the printing press, in the person of the town crier. Scrolls and handwritten books had only limited audiences. In China woodblock printing made possible the distribution of flyers and books to that portion of the population that was literate. Once the printing press was invented and its more automated spin-offs were developed, it changed the face of the distribution of information. It created a passionate desire in everyone to become literate and to be able to read some of the millions of books that were being printed. It also made it realistic to dream of creating a book to be printed. Plus, of course, as more people became literate, the potential audience for books widened, a self-reinforcing process. As printing presses became faster and documents became easier and cheaper to create, newspapers and magazines became very popular, and professional journals proliferated. All this provided a huge boost to innovators because important information was readily available. The more recently invented radio, television, computers, and internet all added to the volume, speed, and accessibility of information.

Not only has the distribution of information been important for the growth of innovative activity, but it has played a crucial role in the spread of political ideas. Newspapers and news magazines in particular were hugely democratizing innovations because they enabled dissenting voices to publish their opinions and versions of important events to counter the official pronouncements, to the point that the press is now considered to be an informal "fourth branch" of government. News periodicals also enabled people to effectively organize and form political parties. The successes of one country were immediately noticed by neighboring countries. New inventions created a demand by merchants, businessmen, and monarchs to have the same capabilities. New institutions like patent offices and limited liability companies were soon emulated as their benefits were recognized. As new markets were developed and profits were being made in one country, businessmen in other countries soon learned about them and wanted a piece of the action. The widespread availability of information was the main factor behind the breakup of the Soviet Union; the Communist party was not ultimately able to keep the people from seeing the prosperity and living styles of the West and from contrasting them with their own situation. Dictators of today attempt to limit the access of information about the outside world to their citizens; they realize that knowledge is power, and that is very threatening to them.

The democratizing impact of advances in communications is less obvious, but it is still powerful. "Communications" as used here differs from the distribution of information; it refers to direct communications between people: person to person or group to group. In engineering terms, it involves point-to-point rather than point-to-multipoint messaging. In ancient times, long-distance communications were mostly between rulers or their agents (such as their military leaders and merchants), and couriers carried their communications. In the Middle Ages, ship- and horse-based courier services were established for carrying messages between monarchs, between noble families, between merchants, and between branches of banks and other businesses. The Tasso family initiated such a service in the 1400s in what is now Italy, and in 1504 the family was given a monopoly by the Holy Roman Emperor to establish mail service throughout the empire.[9] Mail service expanded throughout Europe and enabled scholars and thinkers like Galileo and Kepler to communicate their ideas with each other. This type of service persisted into the 1800s, the pony express being an example well-known to Americans. Since then, trains, trucks, ships, and airplanes have carried most of the mail around the world. The telegraph was considered to be an almost miraculous invention when it was introduced. It immediately became an important tool for conducting business and transmitting news stories around the world. Soon afterward the telephone was invented, and telephone services replaced the telegraph for most communications. After the invention of the telephone-based facsimile, or fax, relatively large documents could be transmitted more efficiently than by telegraph.

The early mail services were too expensive for most people, but by the 1800s many people could attach stamps to letters and packages for delivery, making it possible to communicate with relatives. While telegrams and telephone service were affordable only by relatively rich people at first, the costs eventually came down to where ordinary people could use them. Today cell phones seem to be ubiquitous; these, along with internet-based communications, have proved to be important in exchanging information and getting information out of countries where dictators and other governments have been able to control most of the channels of distribution.

Entertainment, which is taken here to include any leisure activity pursued for pleasure or fulfillment, does not usually come to mind as a revolutionary type of innovation, but it is. In the minds of the ancient autocrats, ordinary people were primarily considered to be sources of labor and taxes. They were there to form the mass audiences when the rulers wanted to show off their power at parades, public spectacles, religious feasts, and holidays. Hence, the rulers used every opportunity they could to intimidate them and to limit their mobility and power. So when the Greeks built amphitheaters and stadiums for the amusement and entertainment of the citizens, they opened an important door for the development of democracy. These innovations were revolutionary because they carried the important message that citizens had the right to enjoy leisure time (i.e., that their wants and desires mattered). Since then, few autocrats and dictators have eliminated sources of entertainment even as they continued to try to control and use them to distract the populace, hoping to limit them to being a harmless energy outlet. Since the Greeks, entertainment has continued to be popular. In Shakespeare's time, some cities had theaters that held plays. The Globe Theater in London had seats for the well-to-do, but it also had areas where the less wealthy could stand and watch the plays. Even before the printing press, books containing stories like those about knights and princesses were popular in Europe, and similar books were popular in China. By the eighteenth century, novels were widely popular, although some religious authorities and government officials actively censored many of them—and still do so today. With the invention of records and movies, followed by radio broadcasts of music and stories, the people's right to be entertained was firmly established. Television, cable networks, and the internet have now made this entitlement irreversible. Concerts, books, athletic games, and theatrical events that were once available only to the wealthy are now accessible to almost everyone. By now, the concept of entertainment has expanded to include other uses of leisure time, such as eating out, tourism, and even shopping. Today it is difficult to believe that there was ever a time when people were denied the right to the entertainment of their choice.

How the Elements Worked Together to Create an Explosion of Innovation

This chapter has described a number of elements that were crucial to the explosion of innovation that took place since about the seventeenth century in certain key places of the world. These elements worked together to reinforce each other and to form a virtuous set of feedback loops:

- *Empowerment*: As people felt more empowered, they recognized that they benefited from increased knowledge. Many more of them recognized that they, too, might be able to invent some useful things or develop useful ideas, so innovation became more highly valued. They also recognized that they didn't have to be stuck working for someone else and doing their bidding, but they could strike out on their own; this had the effect of encouraging a market economy. The newly empowered people passed on these feelings to their children, who felt even more empowered.

- *Expansion of knowledge*: As people had access to more knowledge, they realized that they didn't have to rely on the interpretations of experts, so they felt more empowered. It also made people realize the incredible complexity of nature and life; while this was and still is overwhelming for many people, for innovative people it was perceived as a source of opportunity. These opportunities in turn expanded the market for ideas and products. As knowledge expanded, the demand for even more of it increased.

- *Perception of innovation*: Once people in power accepted the notion that the changes caused by innovation greatly outweighed the associated problems, they rewarded innovative people. This perception also created a demand for more knowledge to support inventions and innovation, and it supported the expansion of competition and a market economy.

- *Emergence of the modern market*: As monopolies were replaced with competition, a large number of merchants and others saw opportunities to make money by making and selling new or improved products, or by finding ways to make them cheaper and thus undercut the competition. Artisans and craftsmen could participate, and with the establishment of the limited liability company, there were new methods of obtaining financial support, and these were available even to less wealthy people. The benefits of knowledge and innovation became more widely appreciated.

- *Democratizing innovations*: As transportation, information, communication, and entertainment became available to the population at large, people felt greatly empowered. They became more literate and better informed politically, traveled to see interesting places and met new people, learned more about all aspects of life, talked to a wider variety of people, and experienced the joys of having leisure time and being entertained (a pleasure previously available only to a privileged few). They became more willing to move to areas away from home to pursue better work. Instead of resisting change, they welcomed the new inventions: electricity, lights, and appliances; cars, air travel, GPS navigators, telegrams, radio, television, telephones, cell phones, records, movies, cassette tapes, CDs, DVDs, computers, laptops, tablets, and multipurpose phones; new medical devices and techniques; checking accounts and ATMs; and literally thousands of others.

John Kay, in describing the workings of the market economy, describes this process as a "co-evolution of economic institutions, social developments, and technological innovation."[10] He says "there is no linear cause; each strand of development both supports and requires the others." That description applies here as well.

Having seen the innovations that took place in different parts of the world, in different geographic areas, under different types of governments, in different types of social structures, and having identified many of the conditions that existed where innovation thrived, we are now in a position to examine previous attempts to identify the conditions that support innovative activity. The next chapter will evaluate these previous efforts, and the following chapter will synthesize what has been learned into a set of five essential factors that I believe best describe the conditions that enable the innovative spirit to thrive.

8. The Conditions that Support Innovation—Previous Work

Evaluating previously proposed factors; demography, geography, geology; governmental and institutional considerations; societal attitudes toward innovation

As I said in the introduction, a strong motivation for writing this book was that I did not find previous attempts to identify the causes and conditions for innovative activity to be satisfying.

Developing a useful list of factors and conditions is not an easy task. Just the fact that many highly intelligent and informed people have attempted to do so shows that it needs to be approached carefully. Their ideas can't be dismissed out of hand. In order to reasonably critique both old and new ideas, we need both a good definition of innovative activity and a set of criteria for the conditions we are evaluating.

To repeat the definition provided in the introductory chapter, the innovations that are addressed in this book are devices, procedures, concepts, theories, or organizational structures that are justifiably considered to be new and either useful or descriptive of the physical world. Utility and the natural world are key distinguishing features.

How does one evaluate the conditions that have been offered up as encouraging and supporting innovative activity? It is easy to come up with a large list of conditions and factors that exist in innovative societies, but which ones are really relevant? Take tolerance as an example. We see more tolerance in innovative societies, but is tolerance a cause of innovation, or are both tolerance and innovation made possible by other conditions?

To help answer these questions, I will use the following guidelines to provide some focus in the evaluation of the various proposed conditions that cause, support, and encourage innovation:

1. Societies where the condition is present should be more innovative than those where it is not, other factors being equal; of course, the tricky part of this is that other factors rarely are exactly equal.

2. The more widespread the condition in the society, the more innovative activity it should exhibit; for example, a society in which all adults are given the same opportunity to innovate will outperform one in which only an elite can participate.

3. If the condition is nearly nonexistent in the society, there will be little innovative activity, regardless of other factors. In other words, it should be a necessary condition.

4. The condition should be applicable to societies over time and throughout the world, rather than to just a few societies and windows of time that may not be typical. It should also be applicable to organizations.

5. Finally, the collection of conditions should be as complete as possible. By knowing how societies perform against the factors in the list, one should be able to accurately predict how one society would rank against others. Then by examining different societies using the list, a report card of sorts that matches estimates of the innovative activity of current and past societies can be developed. Such a list can then comprise the guidelines for encouraging innovative activity, and clarify what defines the innovative society.

Other questions shed light on the usefulness of the proposed conditions:

1. Do the conditions apply equally to both white-collar and blue-collar innovation?

2. Do the conditions stand on their own, or do other factors need to occur concurrently?

3. Are the conditions really features of the society, or are they better described as enabling resources, such as minerals, good soil, or water?

It is also worthwhile spending a little time on the idea of causation, because we like simple models where if one thing happens, then something else will always happen, too. Then we could say that such and such a condition "causes" innovation to occur. That type of thought was what the Greeks and the scholastics like Thomas Aquinas valued. The notion of causation, wherein the existence of a set of conditions inevitably causes a stated outcome, is now considered a quaint model from antiquity. We now know that such cause-effect pairs are rare. Quantum theory shows us that even in the most quantifiable field, that of physics, processes are probabilistic rather than certain.

People are far less predictable than objects, and the physical and cultural environments they live in are very different and highly complex, so that to say that under a given set of conditions innovation will always occur would not be credible. Even the words in English don't always translate well into other languages; for example, the word "innovation" is sometimes used in other languages to mean "heretical," and it has had this connotation even in English. Also, people vary considerably in their innovative abilities and interests. The complexity of life's experiences and the sheer number of human activities virtually guarantee that the simple cause-effect model will be inadequate to describe innovation. So the best one can expect to do is to develop a set of factors that provide a high probability of predicting which societies will be innovative.

At this point I will examine and evaluate nineteen factors that have been cited by various authors or are commonly assumed to be important; fourteen of these are cited by Diamond.[1] The list can roughly be divided into three groups: geographical and demographic conditions, human institutions, and societal attitudes.

The factors in the first group are demographic, geographic, and resource related.

1. *Moderately long life expectancy.* It is true that a long life enables individuals to learn skills and acquire knowledge beyond that required for survival. It also increases the likelihood that people will have the necessary time for reflection that will enable them to identify problems and develop solutions. Particularly in modern times, it takes years beyond childhood to acquire the knowledge necessary for significant innovations, whether blue- or white-collar. Once the general population achieves a life expectancy of about fifty years or more, there appears to be no evidence that longer life expectancy is associated with greater innovative activity. Furthermore, there are many non-innovative societies today that have reasonably long life expectancies, so I conclude that long life expectancy is not a useful factor.

2. *A sufficient concentration of people.* It has often been asserted that cities are more supportive of innovative activity than rural areas. The presence of many other people in towns and cities often means a wide variety of experiences, cultures, and opinions, all of which stimulate thinking people. With towns and cities come larger markets for new products: a craftsman who invents a useful tool or even a novel toy can hope to sell more in a town than in a small village, or even several nearby small villages. The anonymity afforded by cities reduces the limits often placed on individuals in towns and villages. By contrast, rural areas are more likely to be less stimulating, particularly for white-collar innovation. There is no question that contact with lots of different

people can stimulate ideas, but some cities are far more innovative than others. As we shall see, more relevant are the institutions that cities develop (i.e., institutions that provide knowledge, training, and meeting-places for people). A strong argument against this factor is that in many rural areas there has been considerable blue-collar innovative activity; it just hasn't received the same attention. Hence it appears that the concentration of people is best considered as an enabling factor for some kinds of innovation, not as a necessary condition. This will be discussed further in chapter 17.

3. *A favorable geography.* Certainly the geography of a country or society has a major effect. Athens and other innovative Greek cities had harbors, which meant that they had considerable contact and trade with other societies. Sparta, on the other hand, was landlocked and reclusive. However, it's the intercultural contact that is significant to innovative activity. The harbors and roads, ships and wagons, and trade between countries are just the means for making the contact possible.

   The relative isolation afforded to different European regions protected them from being unified by a single autocrat, a point emphasized by Diamond.[2] The geography of China, on the other hand, made it possible to be ruled by a single emperor. The European societies were close enough together that there was a lot of contact between them by road and by sea by the late Middle Ages; this enabled the spread of new ideas and inventions.

   It is obvious that the agricultural potential of some countries is much better than others. The availability of water and other resources differs significantly. However, the effect is more significant in terms of the size of population that can be sustained rather than in terms of innovative activity.

4. *A favorable climate.* Because Europe and the United States have been centers of innovation for a few hundred years, there has been a temptation to attribute this at least in part to the temperate weather. It could be argued that in places like India, the hot weather is enervating, and that in places like Africa, diseases, parasites, and lack of arable land have inhibited innovation. On the other hand, Egypt and Mesopotamia had relatively hot climates but had considerable innovative activity. Landes[3] presents an excellent description of what the effects of an unfavorable climate have been on the motivation of human beings. He concludes that

while it can't be ignored, it is not ultimately a limiting factor. Today, with new technologies providing cooler working conditions and reduced susceptibility to diseases, there is no reason why innovative activity can't be just as vibrant in hot regions.

5. *Abundance of natural resources.* What probably motivated scholars to cite natural resources as a factor was the seemingly reasonable assumption that the more natural resources there were, the more opportunities there were for inventive people. A potter with access to thirty glazes would indeed be likely to create a wider variety of patterns than one who had only five glazes. However, this confuses innovation with variety and artistry. In fact, the lack of natural resources is often a spur to innovation. Contrast the high innovative activity of Israel and Japan, which have few natural resources, with the relatively low activity of Russia, which has lots of natural resources, during the latter part of the twentieth century. This leads me to conclude that other factors in the cultures, not the natural resources (including human resources like slaves), are what make the difference.

6. *Nonavailability of cheap labor.* Paradoxically, the lack of human resources has also been cited. The absence of cheap labor has been said to be a driving force in the early blue-collar agricultural innovations in the United States. The rise of the cost of labor was a motivating force behind the rise of productivity in the 1900s, when factory operators strove to replace workers with machines where they could. The motivation to replace labor by machines in order to reduce production costs continues today to encourage this kind of innovation. However, labor is just one of many resources for which this is true. As oil, gas, and minerals become scarce, the necessity to find substitutes for them similarly motivates innovation. So it would be more accurate and complete to generalize this idea to all resources. When resources are in use and they become scarce, the need arises to find less expensive substitutes or alternative techniques, or to reduce demand for them. However, the nonavailability of cheap labor doesn't meet the criterion of being a necessary condition because innovation can flourish where labor is cheap.

7. *Intercultural contacts.* Intercultural contact stands out as a strong candidate among the proposed factors. Isolation breeds superstition and distrust of outsiders and promotes conformity and

the fear of differences. Isolated tribes, villages, and regions tend to have the most parochial of religions, for example. Worst of all, isolation limits exposure to new ideas and information.

From the middle of the first millennium BC, cities like Ephesus, Pergamon, Thebes, and Athens were continually in contact with traders from all over that part of the world. These traders brought new inventions with them, along with new knowledge of practical techniques, new materials, and new ideas of how to use them to make life more interesting, beautiful, and laborsaving. They learned about Egyptian astronomy and geometry, Egyptian techniques for cutting and shaping hard materials, and the Phoenician alphabet. Sparta, by contrast, saw little of this, remained isolated, and produced few innovations.

The early Islamic period was distinguished by trade and contact with China, India, Persia, and Egypt. Scholars collected books and scrolls from all over the world, including ancient Greek writings. They then assimilated this new knowledge to produce innovations in mathematics, practical astronomy, and navigation.

Intercultural contact has been and continues to be an important stimulus for innovative activity, and its opposite—isolation—discourages innovative activity. So a policy of increasing intercultural contact would definitely encourage innovation. However, it is but one type of access to information, which I will claim is one of the essential conditions for innovative activity; it is discussed in chapter 11.

Except for intercultural contacts, none of the above conditions is useful in distinguishing innovative societies from those that are not. Some do enable innovative activity to occur, but they don't satisfy the conditions that were set out above.

The next group relates to governmental and institutional conditions. Whereas the ones discussed previously are related to geographic location and demographics, the following conditions are created by human activity.

1. *A centralized government.* Central governments have been with us since the time when agriculture supported large population densities. A central government reflects and defines what is important in the society, which in many examples so far have been determined by an individual ruler. A central government provides order, stability, education, and monetary support. When

innovations do occur, the central government performs the role of recognizing them and putting them to use. A central government can reward innovative activity indirectly through its legal structure (e.g., patents) and directly through monetary awards. Most regions of the world now have centralized governments, but their societies vary widely in the extent of their innovative activities. So it's not the fact that there is a central government that matters, but rather it's the nature of the governmental institutions and other conditions within the society. Government can play an important role, often by imposing less control rather than more, but its policies are more important than its centrality. Thus a centralized government is not a meaningful condition.

2. *A stable society.* The argument for stability is that it is difficult for innovation to flourish in an unstable society, which is a valid statement. If a society is frequently being invaded by rapacious warriors or if the government is being overthrown every year, it is hard for people to invest their time and resources to endeavors that may be destroyed or confiscated. Under such conditions, people are more concerned with survival and retaining what they have. Stability provides time to reflect and to experiment. However, it's not true that increasing stability means increasing innovative activity—in fact, too much stability can cause stagnancy, or a total acceptance of the status quo. A certain amount of stability is beneficial, but excessive pursuit of stability can lead to a society with a rigid class structure, one that resists change. As with medicines, too little is ineffectual, and too much can be harmful. Innovation can only thrive where change, at least certain types of change, are welcomed and encouraged. So stability doesn't meet the requirements for being a useful factor.

3. *A favorable class structure.* This condition is difficult to pin down, not the least because people differ significantly in what they consider favorable. Even today there is a widespread disagreement on the most desirable amount of inequality in a society. Presumably the historians citing this were thinking that "favorable" means a relatively weak class structure where people are not bound by birth to live their lives in one class. A rigid class structure reduces the incentive for people to seek ways to improve their lives. It also carries the message that change is not good, that it might threaten the societal structure. It is noteworthy that the great advances in mathematics in India took place prior to the time when the caste system became imbedded in the society. On the

other hand, as we have seen, autocracies did support certain types of innovation within relatively rigid class structures. A major problem with this proposed factor is that it lacks specificity in the sense that a favorable class structure would provide the entire population with the proper education, would make sure that everyone has access to the appropriate information, and would reward everyone for their efforts, which is valid; but it's really these conditions that are relevant and meaningful, not the class structure itself. These considerations are addressed in chapters 11–13.

4. *Opportunities for technical training.* Of all the factors considered so far, this one comes closest to meeting all the requirements I set out. The more people who are trained, the higher the likelihood that innovative activity will occur. Without such opportunities innovative activity is limited to knowledge gained from local lore, which was the case for blue-collar efforts for most of human history. Training is certainly an important factor and shouldn't be limited just to technical subjects, but extended to education in general. It's essential to have such opportunities, but what will make one society more innovative than another will depend on the quality of the training, the quality of the educational system, the quality of the teachers, the fraction of the population that has access to the training, whether knowledge is taught as unchallengeable doctrine or as open to reassessment and improvement, and many other similar considerations. This is treated in detail in chapter 12, which deals with the role of education and mentoring in an innovative society.

5. *Property laws that protect ownership rights.* This is a factor that is primarily applicable to commercial innovations in modern societies, and it certainly affects both white-collar and blue-collar innovations today. What's the point of working hard to come up with a new idea if someone else can immediately copy it, particularly if you are hoping to profit financially? However, such rights were established only in the last few hundred years, whereas innovation has been with us far longer. People in the past were still motivated to come up with new concepts and inventions. They were motivated by recognition from the autocrat, by recognition among their peers, by the sheer joy of creation, or by the pleasure of making their friends' and neighbors' lives easier, not by direct monetary gain. While property laws play an important role in commercial innovation today, they are part of a larger set of

societal conditions that encourage innovative activity. These are discussed in chapter 13.

6. *A supportive economic system.* A supportive economic system makes it potentially rewarding to invest capital in technological development. Like property laws, this factor is specific to commercial endeavors today. However, it is indirect, rather than direct; that is, it does not directly motivate the innovators, who often don't have the capital. Rather, it motivates people with capital to seek out innovative people. It comes under the heading of the value placed on commercial innovation by the society. It is addressed in chapter 13, which discusses societal rewards for innovation. Viewed historically, however, it is certainly not a necessary condition, because as was shown here, considerable innovation took place in autocratic societies where the inventors had no way of collecting royalties—the hope would have been to receive some favor from the ruler or one of the ruling families. Similarly, the monks of the monastic estates of Europe were not motivated by profit, but by other ideals.

7. *A supportive religion.* Religion reflects a societal attitude, but one in which the religious tradition plays a dominant role. Clearly, religion per se is not an inhibiting factor to innovation, and indeed can be supportive. Most religions don't directly inhibit blue-collar innovations, for example. We saw that the monastic estates of the Middle Ages supported many blue-collar inventions, including those used in construction such as pulleys, gears, and ropes. We know that the Shakers in the early American states were moved by their love of God and community to invent and improve upon a number of tools and materials. Individual scientists like Isaac Newton and Michael Faraday had strong religious beliefs and considered their research to be an expression of their religious striving. In the artistic sphere, sculptors, architects, painters, and musicians have similarly used their religious motivations to create beautiful works of art.

   Suppression of innovation has occurred when religious control has been so extensive and pervasive that religious leaders have attempted to enforce a particular view of historical events or of the physical world and have ignored or suppressed contrary objective evidence. Unfortunately, the followers and leaders of monotheistic religions have often fallen into this trap, and they have indeed often impeded scientific progress. Innovation has been inhibited by inculcating fear into the populace, limiting education only to the clergy, restricting access to information, and reducing opportunities for creative discussion of differing opinions. Sometimes this suppression has been in the form of religious sanctions (e.g., excommunication); it has also taken more blatant forms when the government and religion are allied, such as occurred with the Inquisition in Italy and Spain. But often suppression has been less obvious and less dramatic, occurring through social disapproval and ostracism, especially in smaller religious communities, rather than through a remote central religious authority.

While religion can be a motivating factor, it is just one among many others, such as monetary rewards, social approval, fame, and personal satisfaction. Religious authorities can suppress the inventive spirit, and religion can be such a large part of everyone's lives that practical inventions are considered unimportant; but a supportive religion can provide additional motivation for some individuals.

8. *War*. War—or, more accurately, preparation for war—has often created a reward structure for innovation. No doubt warriors and officers who developed improvements in weapons, training programs, siege techniques, or vehicle designs were encouraged and rewarded by their efforts. It is also likely that some of these improvements resulted in spin-off techniques that were beneficial in civilian applications. For example, improvements in chariot designs were no doubt used in popular chariot races and probably found their way into agricultural applications. Certainly the experience of the competing powers in the world wars of the twentieth century described in chapter 6 shows that tremendous technical advances took place in weaponry, aviation, radar, and navigation. While war preparations can result in technological advances with civilian spin-offs, it comes at a very high cost in terms of the society's resources. The highly centralized governments pumped huge resources into war-related technologies during World War II and produced some very impressive innovative activity. However, it was after the war was over that the technological explosion took place. As the industries became increasingly decentralized, the innovative energies of the newly educated populace were released. The wartime developments in radar, navigation, and communications formed the basic science of an explosion of technology that resulted in transistors, printed

circuits, GPS, satellite communications, cell phones, transponders, personal computers, and the internet. By the 1990s, civilian technology development had surpassed that of the military, and today the military incorporates civilian technology into its equipment; technology transfer is primarily from civil to military.

Since the point of the weapons and siege techniques is destruction, not construction, there is little doubt that war has destroyed many inventions of the enemy of the moment and kept them from being widely used by other societies. So war can also destroy innovations.

While preparation for war has stimulated innovative activity in the past, the case cannot be made that promoting war also promotes innovation, except perhaps in military technology. Just as autocracies supported certain types of innovative activity, so has the preparation for war. However, advocating more autocracy or war to encourage innovative activity makes no sense.

A third group of proposed conditions reflects societal attitudes. A societal attitude is one that is held by most influential individuals within the society. Given the wide range of attitudes on any subject within a population, unanimity is unlikely to be found on any issue. However, it is generally understood to mean that these attitudes are described in a positive light in textbooks, newspapers, magazines, and other media.

1. *A societal attitude that values individual efforts.* The idea is that people will be more likely to be motivated to spend their time on inventive activities if they get a tangible benefit from doing so, such as money, a job, fame, or recognition from their peers or from people in power. This attitude got its start with the Greek notion of the individual achiever, where athletes, playwrights, teachers, and authors worked hard to receive acclaim from their fellow citizens. In autocracies the credit for individual accomplishments almost always went to the ruler or to favorites of the court. In collectivist societies most of the credit went to "the people," and the names of the individuals involved were de-emphasized.

   Many people think of innovative individuals as lone inventors working in solitude, but such cases are rare; inventors need access to knowledge and to people with relevant knowledge. Also, today, much innovative activity involves groups of individuals working together. There are many examples of highly innovative groups, such as in research institutes; Japanese carmakers and electronics manufacturers are some contemporary examples of organizations that use groups of experts to optimize product design. However, it is ultimately individuals who make the key contributions.

So there is definite merit to the idea of valuing individual efforts. It is developed further in chapter 13, which addresses societal rewards for innovation.

2. *A societal attitude that allows and encourages risk taking.* This is similar to the condition for valuing individual efforts. If the risks implied here are financial, then this factor has some applicability to commercial inventions; usually, however, the risks are taken by the entrepreneurs and financiers rather than by the inventors. Many blue-collar innovations or innovations that come out of research and development institutions don't involve very much risk. It is true that some innovators have indeed been fearless risk takers; it is also true that a society that discourages risk taking by inculcating a fear of failure or attaching shame to a failed effort does not provide an environment encouraging to innovation. This topic is addressed in chapter 10.

3. *A societal attitude that values the scientific outlook.* Presumably the scientific outlook cited here means an attitude of respect toward scientific investigations. An example of its opposite would be a highly religious group that believed scientific investigation to be an affront to their deity. But science as a separate activity and the scientific outlook are quite recent concepts, and the associated innovative activity is primarily of the white-collar variety. As a consequence, it is not helpful in assessing earlier societies or blue-collar inventiveness. Even today there are many people who are suspicious of the scientific outlook but who embrace technological advances wholeheartedly. It would be more useful to focus on the societal attitude toward change, where scientific exploration that overturns commonly held viewpoints is either tolerated or resisted.

4. *A societal attitude of tolerance of diverse and dissenting views.* Tolerance and freedom of thought are strongly associated in most people's minds with innovation. Tolerant societies have typically outperformed intolerant ones when it comes to creativity and innovation, so there is definitely a strong correlation between tolerance and innovation. It is also easy to understand why intolerance would have a negative impact on creativity because any deviation from the norm—any change, any suggestion that the

status quo is erroneous or inadequate—would be resisted. Anyone suggesting a better teaching technique, as just one simple example, could be accused of criticizing some current teacher and, by extension, of criticizing teachers in general. An intolerant society would discourage such thinking, while a tolerant society would be more likely to have people in it that would be interested in seeing if the proposed technique had merit. Tolerance for new ideas implies risking criticism of the proponents of current practices.

There are, however, several problems with associating tolerance with an atmosphere friendly to innovation. First of all, many people who have been recognized for their innovative and creative accomplishments were not tolerant people. Innovative people have often been quite difficult and imperious. Second, tolerance in one aspect of life does not always imply tolerance in other aspects; societies that tolerated political differences didn't always tolerate religious differences, and the dominant participants in some societies who considered themselves to be tolerant were often quite intolerant of poor people or of people from other cultures. The bias of European historians of the nineteenth century against the contributions of Africans and South Americans is a notorious example. Third, even if a convincing case could be made, it's not a useful factor because it leaves unspecified how such tolerance can be encouraged or promoted. A more useful approach is to identify what causes diverse and dissenting views to be considered and discussed in a society in the first place. In other words, there's not a causal relationship at work between tolerance and innovation. It is more accurate to say that societal conditions that encourage tolerance usually encourage innovative activity.

A number of proposed conditions are necessary for white-collar innovation, specifically, to thrive. Some of these have already been mentioned above, and others have been cited or implied by historians. They aren't as applicable to blue-collar efforts in the past, but they are applicable to most current innovative activity.

1. *Media for writing.* A written language needs writing materials. A written language has to be recorded using materials like brushes and ink on a medium such as clay, parchment, plaster, stone, papyrus, or paper, one that lasts for a long time and can be conveniently stored.

2. *A large group of people.* There needs to be a reasonably large group of people with the potential for contact with each other and

who share a common language. Then the needs of the society are sufficiently complex that work is highly specialized, so that there are many opportunities for innovative activity. In independent small towns and villages, on the other hand, the predominant needs were often for food, shelter, and protection against marauders—life was simple and hard, and almost everyone spent their time dealing with issues of survival. A large group of people doesn't necessarily imply millions. As we saw, the Greek city-states produced some of the most amazing minds ever with relatively small populations. Athens probably had less than half a million people, typical of the cities at that time.

3. *A centralized government.* Wherever there has been a large group of people, there has been some kind of central government to organize life, so this condition is closely related to the previous one. The central government reflects and defines what is important in the society, which in the examples so far have usually been bestowed to an individual ruler. A centralized government provides order, stability, education, and monetary support. When innovations do occur, an encouraging central government performs the role of recognizing them, putting them to use, and making them known throughout the region.

4. *An educational system.* The society needs a means to train at least a portion of the population to read and understand its recorded knowledge. In this way the knowledge and innovations of previous generations can be reviewed and understood, enabling them to be used and improved upon.

5. *Time for reflection.* Innovation, and indeed all intellectual endeavor, requires that at least some people have time over and above their routine duties to ponder and to examine life and nature. Innovation requires the ability to see things from a perspective different from the conventional. Current knowledge needs to be thoroughly understood and questioned, and this can't happen with people whose days are spent performing routine tasks. Until the last few centuries, this has meant that white-collar innovation was limited to the clergy, the nobility, and members of rulers' courts.

One might claim that these conditions are necessary for innovation to occur, but even if they are, they are not sufficient. That is, even when all these conditions are present in a society, it doesn't guarantee that there will be innovation. If people see no benefit to spending time trying to improve their surroundings, they will not do so. If what is relevant and rewarded is determined by a ruler, then other endeavors will not be attempted. If education is limited to priests and religious matters, then intellectual thought will be primarily religious, and nonreligious activities will be considered to be less relevant. Indeed, looking at the countries of the world today, it is evident that most of them meet all five of these conditions, but only a fraction of them could realistically be considered as innovative. So none of these factors explain why some societies are more innovative than others.

In summary, having considered the list of factors and conditions that historians have suggested, I conclude that some of the factors that have been proposed by various authors are not useful, some of them both encourage and discourage innovation, some are limited to particular kinds of innovation, and some result from other conditions that exist in the society. It is not possible to pick a few of them to form a complete group of factors that collectively enable us to make an overall assessment of innovation in a society. It is better to come up with a synthesis of the good ideas and augment them with others.

The primary motivation for this book was to take a fresh look at the aspects of a society that encourage innovation and to identify a set of supporting conditions that is more causative, in the sense that the presence of these conditions prepares the way for innovative people to do their magic, and supports their efforts. Once identified, the essential conditions can be established or encouraged by policies and movements within a society. By establishing such conditions, societal attitudes that support innovative activity will generally be encouraged, and the opposite attitudes discouraged. Finally, by focusing on establishing these conditions, specific proposals for policies and laws can be evaluated for their potential effectiveness, and ineffective proposals can be avoided. This is the challenge that I take up in the next chapter.

9. The Conditions that Support Innovation—A New Look

Synthesizing the best ideas into four essential factors; adding a new one

The historical discussion presented here has described the what, who, when, and where of inventions and developments, but it hasn't addressed the why and the how in any detail. The previous chapter discussed a number of conditions that have been proposed by historians to address the why and how, and we have seen how they fall short of providing a satisfactory explanation. However, some of the proposed conditions had features that merit further attention, both for blue-collar and for white-collar activity. In this chapter I will draw and expand on those, and I will synthesize these and other ideas into a concise list that incorporates the most important concepts.

Before I describe my proposed set of societal conditions, it is instructive to further examine some of the previously proposed factors as they apply to blue-collar and white-collar innovation. The innovations that took place outside the cities did not depend directly on having a large number of people living close together, which was one of the proposed conditions. Once an area became known to have metal ores, marble, or other useful building materials, for example, the local king would want to exploit the resources for use as military weapons, for building materials, and for trade. A classic example was bronze, which is made from copper and tin, materials not usually found near each other. Copper by itself is a soft metal, but the addition of tin to the tune of about 12 percent makes a bronze alloy that is quite hard. The Phoenician traders went as far as Britain to obtain tin, which they brought back to the Mediterranean. In such a situation a king would have been highly motivated to exploit the natural resources, so his agents would have been looking for ways to locate the resources, to improve the methods of extracting, melting, and purifying the metals, to shape the metal into useful objects, and so forth. Workers would have been rewarded for improvements they came up with. They were illiterate, so writing tools were not relevant to them. So many of the proposed conditions in the previous chapter don't apply well to blue-collar innovation, especially in ancient times. Intercultural contacts certainly played a part, though; it was common for foreign specialist workers to be brought in when new resources were discovered.

White-collar innovation prior to the last few hundred years was motivated not by patents or profits but rather by prestige and paid positions. A societal attitude of tolerance toward risk was not much of a motivator either. The positions in government, being political, depended on contacts and good will as well as talent, so the safest lines of endeavor were those that supported the ruler and the religion (such as astrology, astronomical predictions, medicine, or surveying techniques) or that would not be likely to offend anyone (such as highly abstract endeavors, like mathematics). The Greek thinkers were allowed more freedom, but even they were considered suspect by many of their more religious or patriotic fellow citizens; Socrates was sentenced to death apparently because he opposed the concept of "might makes right." In the medieval Italian city-states, it was fashionable among the wealthy to sponsor scholars and artists; Galileo got Cosimo Medici to offer him a position in Florence. In such circumstances thinkers were much freer to explore new ideas and conduct research on a wide variety of topics related to the natural world. After the Reformation, this freedom became even more widespread, and many more talented thinkers were attracted to what we now call scientific endeavors. While this picture fits well with the notion of a societal attitude that values individual efforts, it doesn't fit so well with a notion like a supportive religion or economic system—the thinkers were tolerated rather than supported. Most of them were reasonably well-off, so the class structure wasn't highly important. On the other hand, their efforts depended strongly on being well educated and having access to the wisdom (and follies) of the ancients, as well as those of other cultures.

In the last couple of hundred years as the blue-collar/white-collar distinction has become less relevant, economic considerations have become more important. Patents, research positions, and entrepreneurial opportunities all encourage innovation at the citizen level, while governments sponsor research in military and infrastructure applications. Information has become readily available worldwide. Many of the factors discussed, like geography, climate, availability of natural resources, and the concentration of people, are no longer as relevant as they might have been in earlier times.

When one speaks of "the society," it is necessary to recognize that few societies are homogeneous, and cultural values are rarely unanimously held. We all live in, belong to, and are affected by a wide variety of subcultures: different religions or denominations, different economic circumstances, different amounts of education, different families, different age groups, different national backgrounds, and different peer groups. Today many of us live in societies with many ethnic groups, so these differences are more apparent to us, but in fact this plurality of subcultures has been the experience of most of the human population throughout history. To speak of the culture of a country is almost always a gross oversimplification. People in cities have different attitudes from people in the country. People born into wealth have different attitudes from people born into families that have to work to survive. Immigrants have different attitudes from people whose families have lived in the area for generations. People in different occupations, trades, and professions have different values. These differences are reflected in a wide variety of attitudes toward change and toward new developments and new ideas.

Still, while many of subcultures are present within a society, there is usually a dominant culture that everyone recognizes, one that is reflected in the government, laws, societal myths and histories, family ties, and religions. If government officials and the leaders of business, education, and religion who represent the dominant culture are threatened by new ideas or changes, then innovation by individuals, no matter how highly motivated, will be discouraged. Similarly, even if the dominant culture is favorable to innovation, it may view subcultures within the society unfavorably, which will suppress innovative activity within those groups. The converse is also true, in that pockets of innovative activity may be able to thrive even if the dominant culture is hostile.

Some writers of history have fallen prey to oversimplifications and have made statements suggesting that a society "produced" some individual innovators or that "the time was ripe," subtly undermining the individual contributions of the innovators. Such a deterministic model makes no sense—there is no inevitable cause-and-effect mechanism at work. It is more accurate to state that when certain conditions are present in a society, it is more likely that individuals will exercise their innovative skills.

The opposite oversimplification is also invalid—namely, that the history of innovation is basically the story of a few individual geniuses. Such an assertion would imply that blue-collar innovations were inconsequential or irrelevant, and this is manifestly not the case. It would also imply that inventions were developed occasionally and suddenly in their final form, when in fact virtually all inventions were the result of many small improvements followed by flashes of inventiveness that required intelligence and diligence, but not blinding genius. These were then followed by further improvements that made them more useful. If one insists on calling such inventors geniuses, then they were not few, but many; there have been tens of thousands of people who have performed significantly innovative work. As Conner points out,[1] most scientific advances have been built on the work of a large number of what I call blue-collar innovations. The genius hypothesis also flies in the face of the fact that a significant number of inventions and innovations came about by people attempting to solve problems totally different from the eventual applications of their work; these efforts were not the works of masterminds who recognized a problem and formulated a concise solution. Reality is much messier and much more interesting. Other writers, like Rosen,[2] have made this argument. Rosen also cites Usher,[3] whom he noted hated both the "great inventor" theory (which claims that the history of innovation is the work of a few geniuses) and the "societal production" theory (which sees innovation as the consequence of "immutable laws of history"). He firmly believed that the ability to invent was present in many ordinary people and that the impulse to invent was universal. The books of the three authors mentioned here provide a host of examples to back up these assertions.

This book takes it as self-evident that human beings have the innate ability to innovate and create and that different people have these abilities in different degrees. Thus if the conditions are right, it is highly likely, but not inevitable, that some individuals will exercise their abilities to innovate. This fits my own personal experience and my reading of the historical record.

But what does it mean to say "the conditions are right"? What are the factors that distinguish innovative societies from the others? The conditions described in the previous chapter are relevant to the story of innovation, and some are quite significant, but the purpose of this book is to develop a brief, coherent set of conditions that synthesize and distill the best ideas. A well-formed set of factors would provide a scorecard of sorts that would make it possible to assess government policies, corporate policies, and cultural attitudes, a list that would suggest actions that could be taken to improve the innovative environment. This group of factors should apply to old and new societies everywhere and should apply to both blue-collar and white-collar innovation. Ambitious? Perhaps, but it can be done.

In chapter 6 it was shown that innovative activity surged when people felt more empowered, available knowledge expanded, a competitive market developed, and people across the society accepted innovation as desirable and rewarded inventors. These considerations form the basis of the conditions required for an innovative society.

Recall the criteria cited in the previous chapter that were used to evaluate each proposed condition supporting and encouraging innovation:

1. Societies where the condition is present should be more innovative than those where it is not.
2. The more widespread the condition in the society, the more innovative activity it should exhibit.
3. It should be a necessary condition (i.e., without it there would be little innovation).
4. The condition should be applicable to all societies, whenever and wherever they existed.
5. The collection of conditions should be as complete as possible.

Stated another way, the conditions should be individually necessary, and collectively sufficient, to the extent possible.

The next five chapters describe in detail the five conditions that I claim are definitive in distinguishing an innovative society from one that is not. All of these conditions were present in the countries where innovation thrived in the last two hundred years; less innovative societies lacked at least one, or were deficient in several of them. The five major conditions are the following:

1. *Absence of fear.* The absence of fear results in the empowerment of individuals. If people feel that they can develop new processes, gadgets, or concepts and talk about them openly and bring them to fruition, then the innovative ones among us will attempt to do just that. If, on the other hand, these same individuals believe that their efforts will offend the community (or local government officials or their bosses or some other powerful person) or that they might lose their savings and homes or reputations, then they will decide it's not worth the effort.

   It is evident that the absence of fear meets the criteria listed above. Fear is more prominent in non-innovative societies. The less fear in the society, the more innovative it is. Where fear is rampant, little innovation takes place. This is true for societies in the past and present throughout the world.

2. *Access to information.* This factor not only refers to the means and convenience of access to information, but also to the relevance of the information, and—even more important—to the number of people within a society who have this access. If only a select group like a priesthood or emperor's court have such access, the

innovative energies of the rest of the population will not be harnessed, to the detriment of all.

This condition also meets the cited criteria. Societies with more access to information are more innovative than those with less, and the more people within the society that have this access the better. With little access to information, there is little innovative activity. This has been and continues to be true everywhere.

3. *Guidance in the use of information.* Certainly this factor involves the general education of the populace, but it also includes the quality of the education, the extent to which it is available to all, and the accessibility of education. In addition, it includes the mentoring that takes place within a subculture, such as occurred in the guilds. Furthermore, it is important to assess education in terms of its relevance to innovative activities and in terms of the guidance in utilizing information—critical and clear thinking processes must be effectively taught.

   Societies that educate their people and provide appropriate training outperform those that do not, and the higher the percentage of the population participating, the more innovative the society. Where little training in the use of information occurs, innovation is limited; this has been true everywhere over time.

4. *Societal rewards for innovation.* An innovative society will value innovation and reward attempts by its people to introduce new ideas and concepts, even when they fly in the face of accepted notions and practices. Rewards can be in the form of employment, money, opportunities to form a business to build and sell invented products, patents, status within an institution, or just the respect of a few peers. The laws and institutions of the society will provide encouragement and protection for innovative activities and will discourage those who try to restrict them. This concept applies within individual companies and organizations as well as within governments.

   A key feature of innovative societies is that they encourage individual efforts at every level (i.e., through government policies, institutions, companies, and a general cultural interest in progress). This encouragement both reflects and reinforces a positive attitude within the population; the more widespread the rewards, the more innovative the activity. Where rewards are limited, there is little innovative activity. Again, this holds true historically and throughout the world.

5. *A multiplicity of opportunities*. This factor has been largely ignored by historians, but it is one of the most important conditions for sustained innovation. The innovative society will provide a number of choices for education, for the dissemination of new ideas, for the evaluation of new innovative proposals, and for funding new developments. Decentralization and multiplicity are key ingredients. As I will show, over-centralization tends to lead to uniformity and the discouragement of innovation. I will also show that attempts to create a single "best" institution or organization will actually limit innovative activity because life is too complex and rich to be tied down to a single best of anything worthwhile.

   A potential inventor in an autocracy might well have only one potential customer, the autocrat, and if an invention doesn't please that person or his agents, the game is over. In countries where inventors may sell their ideas to any of several possible customers or investors, there is more likelihood of success; as the likelihood of success increases, the more options there are. This multiplicity of opportunities leads to competition, and competition is a huge motivator for innovative activity. In the past, multiplicity just happened. Modern innovative societies deliberately attempt to maximize the number of such choices.

These five conditions synthesize the best ideas from the list of proposed factors cited in the previous chapter, and they incorporate the essential conditions for innovation. They apply to both blue-collar and white-collar innovation. They have the advantage of being tangible, in the sense that they can be used as the basis for the evaluation of specific proposals to encourage innovative activity within a society. Taken together, they meet the criterion of completeness, in part because they cover all the factors and influences that have been identified, but also because collectively they are sufficient: societies that have all five are indeed highly innovative. They also form the criteria for a scorecard for evaluating societies and ranking them for their support of innovative activity. The next five chapters will explore these conditions in detail.

10. Absence of Fear

Fear of the autocrat; fear of powerful people; class-based fear; fear of loss of income; religious fear; fear of ridicule; fear of failure; overcoming them all

The Roman writer Pliny tells the story of a glazier who had invented a process for making virtually unbreakable glass objects.[1] He showed a vessel to Tiberius, the Roman emperor, hoping to sell him objects of this material. After making sure the secret of the process was not known to anyone else, the emperor had him executed, because such glassware would have competed with traditional gold and silver cups and similar articles. An environment like this where fear is deliberately promulgated is hardly conducive to innovation.

Once the agricultural revolution took place, most of the cities and countries were headed by autocrats, who had the power of life or death over the people. The primary concerns of these leaders were maintaining internal control, defending against invaders, invading neighboring countries, and creating visible symbols of power. Internal discipline was maintained primarily through force, which required most of the inhabitants to have a healthy fear of the leaders. The inhabitants were frequently made aware of the presence of those who enforced the laws, and great buildings and squares were built to serve as a constant reminder of the autocrat's greatness and power. In other words, while the structures were impressive and often beautiful, they also served to intimidate the populace.

In ancient times it was common for the autocrats to use religion to increase intimidation by associating themselves with the society's religion and to claim that they were appointed by the gods. Such conditions subtly sent the message that changing the status quo in any way might offend the heavens. Often the emperors or kings would claim to be gods themselves. Priests, who were supported financially by the autocrats, were expected to support this useful fiction. The autocrats were the final arbiters of what constituted acceptable behavior. In such an atmosphere, activities that supported or enhanced the interests of the autocrats were encouraged, but virtually nothing else. This is why the innovations under different autocracies were used almost entirely for controlling the populace, supporting military adventures, and constructing and decorating buildings and tombs.

So even in societies that might have otherwise had good conditions for innovation, this intimidation acted as a deterrent to innovative activity performed by anyone except those who served the rulers. Most of the population lived in considerable fear because they were considered dispensable by the ruler and could be punished or killed if they offended the wrong people. Because they lived at the pleasure of the rulers, who considered them as subjects who were on earth primarily to serve the royal concerns, they did not usually have a lot of time on their hands. The few who were in a position to be allowed the luxury of toying with new ideas were the priests, astrologers, architects, builders, generals, and scribes—the elite who were closest to the autocrats. Priests were allowed to consider philosophical questions, as long as they made sure to keep the population in line. Scholars could too, as long as their writings sang the praises of the rulers. Astrologers could come up with theories about the stars and planets, as long as they found evidence of heavenly justification for the emperor's authority. Architects and builders were allowed to build new types of structures, as long as the king got the credit. Generals encouraged their officers to come up with new weapons or military strategies that helped in defensive and offensive military operations. Scribes were allowed to develop efficient new writing techniques, new symbols, or abbreviations. Artists and artisans were paid to create sculptures, paintings, and furniture. Writers and musicians were paid to praise and entertain the nobility. So some types of innovation and artistry were encouraged under autocracies, but under strict limits.

Along with fear of the authorities, people were inhibited by religious fear, especially when the religion was highly centralized and backed by the power of the state. Religious fear was a drag on white-collar innovation, especially when the religion made claims about knowledge of the material world, its history, and the forces that affect the natural world, such as volcanic eruptions, earthquakes, hurricanes, and floods. Once such claims were made, backed up by awesome threats of displeasing the gods or eternal damnation, few individuals were likely to speculate on alternative explanations. Those who did suffered the consequences.

An example in more recent Western history involves the Copernican theory of planetary orbits. The dominant Christian doctrine was that all heavenly bodies rotated around the earth, and any scholars who might say otherwise were subjected to scorn and ridicule by most, but not all, Catholic scholars at the time. Copernicus (1473–1543) was aware of the risks of publishing his theory, which identified the earth as one of several planets that traveled in circular orbits around the sun, so he held off publishing his work until he was on his deathbed. For some time, discussion of his theory was allowed among religious scholars, but with the restriction that it was an interesting hypothesis, not to be confused with reality; Copernicus himself said in his dedication to Pope Paul III that mathematics, not physics, should be the basis for understanding his work. A century later it was put on the list of forbidden books. The monk Giordano Bruno studied it and concluded that, indeed, the Copernican theory described reality, and he advocated this view. For this and other heresies, he was burned at the stake in 1600. Galileo, who was thirty-five at the time, had come to similar conclusions, but he held off pushing them too hard until later when he thought it was safe. When his friend and patron Urban VIII became pope, he again advocated the Copernican theory, buttressed by his observations using the newly invented telescope. These observations included the discovery of moons around Jupiter, a concept that could not be incorporated into the Aristotelian model of celestial spheres espoused by the Church. However, he had misinterpreted the intellectual atmosphere and was interrogated by the Inquisition and excommunicated, lucky not to follow Bruno to the stake.

Religious fear is not just the kind whipped up by centralized religious authorities. It also comes from the religious community at large. For example, Islam is a highly decentralized religion, and Islamic communities have in the past been quite friendly, tolerant, and even encouraging to different concepts and ideas, as pointed out earlier. Even in the early halcyon days of the relatively open Islamic Empire, however, there were a few Islamic communities that were quite threatened by open scholarship and outside ideas, and today both kinds of communities exist. Fear plays a major role in suppressing innovation because of the power of intolerant communities over individuals. There are many Muslims who recognize the problem. The Moroccan scholar Mernissi[2] discusses the various fears that some Islamic people have about the West and urges the Islamic communities not to place limits on freedom of thought. "We cannot continue to stifle the imagination, the freedom to ponder and dream, for that is the locus of invention, the source of wealth in the electronic age!"

The phenomenon of religious fear is widespread, and not just in some Islamic communities. Small, isolated communities have always been susceptible to fears that associate new ideas and even new objects with diabolical activities. In the 1600s the story of Dr. Faust was popular. He was an extremely smart man who found himself increasingly unhappy with the limits of his library, his associates, and his environment. He knew there was more knowledge to be learned, and he ached to have access to it. The devil offered him all that he wanted, the price being his soul after death, to which he agreed. While he was alive, his house was said to contain great marvels and inventions, and flowers bloomed there even in winter. His story became popular as a morality tale: too much knowledge was dangerous, and being overly zealous in the pursuit of knowledge was ultimately against God's will. The story of Adam and Eve partaking of the fruit of the tree of knowledge of good and evil in spite of God's commandment not to do so is used even today by some religious people as a warning against wanting to know too much.

Punishment by autocrats and religious leaders and communities are not the only sources of fears that inhibit innovation. Seagoing ships were captained by tough leaders that had life-and-death power over the sailors. This was understandable given the need for crew discipline and the relative ignorance of some members of the crew, which often included criminals. However, this was translated into shipboard rules that discouraged independent thinking; in the British navy, common sailors were forbidden to try to keep track of the ship's course—only officers were allowed to perform navigation functions. While this practice had a certain logic, it could also prove disastrous. In 1707 Admiral Clowdisley Shovell was returning to England from some military battles and found himself in the fog for several days.[3] He and his officers, after reviewing what navigation data they had, mistakenly placed themselves many miles west of their actual location. An unnamed seaman had been surreptitiously doing his own calculations and, realizing what was happening, took a chance that he might be able to present his case. Unfortunately, he was immediately hanged for insubordination, and his calculations ignored. It turned out that he probably had been correct because four of the fleet's five warships foundered on the rocks near the Scilly Isles shortly thereafter. Most of the sailors died, including Admiral Shovell.

One fear that has been especially inhibiting to white-collar innovation is the fear of disrespecting admired religious or scholarly figures, and thus being ridiculed, criticized, or even ostracized by the establishment. The attitude toward Greek scholars, notably Aristotle, by the Jesuits and other scholars went way beyond respect; it even went beyond veneration and bordered on worship. Galen had a similar reputation in medicine. As I mentioned in chapter 3, Aristotle's students probably would have laughed at the quasi-deity status ascribed to him during the Middle Ages. In China the veneration of elders and the control of scientific, political, and religious thought by the emperors' bureaucracy resulted in a fear on the part of the scholars and citizens of thinking about concepts that might run afoul of the accepted wisdom. According to Landes,[4] in China in the 1600s "would-be modernizers were thwarted, moreover, not only by brittle insecurities but also by the intrigue of a palace milieu where innovations were judged by their consequences for the pecking order. … At all levels, moreover, fear of reprimand (or worse) outweighed the prospect of reward." He quotes a Christian friar, Evariste Huc, who observed that "any man of genius is paralyzed immediately by the thought that his efforts will win him punishment rather than rewards."

By contrast, Britain's Royal Society had the motto "*nullius in verba*" ("on no one's word").[5] Nothing was to be assumed to be true merely because some genius, including Aristotle, had said it. This became the motto of the Age of Enlightenment in the 1700s and the foundation of the scientific method.

So fear of knowledge, fear of change, and fear of offending gods of tradition or conventional wisdom—whether reinforced by powerful religious authorities, by a belief of the religious community in a society, or by the superstitions of a village or church congregation—can exert a powerful inhibiting force on individuals who might otherwise be interested in considering and developing new ideas. In addition, those individuals who take the leap to do so risk anything from subtle disapproval to outright ostracism from their communities.

If fear constrains innovation, and if the autocrats induce fear in the populace, what are the earliest examples of societies that avoided this fate? The prime examples that come to mind, of course, are the Greek city-states from about 500 BC to 300 BC and the Hellenic city-states around the Mediterranean after that. The Greek city-states (some of them located in what are now Turkey and Italy) of Thales, Pergamon, Athens, and Thessalonica and the colony of Syracuse traded with each other and with Egypt and Crete using the Mediterranean Sea for transportation. The eastern despots were far enough away that they only occasionally attempted to conquer the Greeks, usually seeing them as unthreatening and useful for trade. As noted in chapter 3, the power of the early kings in these city-states was diluted, and the power of the aristocracy was checked by the merchants. This situation inhibited the development of an autocracy. Meanwhile, the society was fortunate enough to have several enlightened leaders who developed democratic laws and institutions. In this environment a citizenry developed that did not fear the authorities, people who were quite proud of their newfound freedom. They built coliseums, temples, theaters, and other buildings not to honor powerful leaders but for their own enjoyment and fulfillment. They argued with each other with much less fear of being punished—in fact, playwrights often mocked the aristocrats in their plays. In this environment white-collar innovation, in particular, thrived.

Most of what has been discussed so far applies mostly to white-collar innovation where the new ideas could threaten some existing power structure in the society. But how would fear act to hold back practical inventions and ideas?

One strong inhibition is a result of a class structure: the autocrats and the ruling classes always felt superior to the masses—the peasants or common people—and arranged to pass on their power to their own children. This was accomplished through the laws of the land, through frequent displays of their power by their dress and ceremonies, by living in lavish homes, and by having their agents control the economic lives of the lower classes. The people in the lower classes had their inferior status drummed into them from birth, and they were used to having their lives circumscribed by their "superiors." Under such conditions, it's not surprising that few were motivated to think in terms of tinkering. In spite of this, there have been exceptions, such as slaves in the United States who invented things that were claimed by their owners. In ancient autocracies, blue-collar innovations of some subcultures such as soldiers and mine workers were given some encouragement, as they furthered the interests of the rulers.

Of course, there were people in the upper classes who were capable of invention, and they would not have been so fearful. However, until the last few hundred years (and even then in only a few places and times), upper class people scorned manual efforts of any kind and were contemptuous of the farm workers, mineworkers, craftsmen, farmers, and others who worked with their hands.

A very important change took place in the Middle Ages in Europe, one that has only recently been recognized, and one whose significance can now be more clearly understood. *The monasteries and monastic estates began to encourage and support blue-collar innovation.*[6,7] This was able to happen for three very important reasons: (1) the lifestyle was not based on class, (2) the tangible inventions and devices that emerged did not challenge the (white-collar) teachings of the religious authorities, and (3) the monks did not attempt to establish monopolies on the production of their inventions since they were motivated by religious pursuits.

The monks banded together for security and community and attempted to create an environment where they could spend most of their time in prayer, reading, and religious observances. However, they had to eat, and that required a lot of time spent raising crops, animals, and poultry. Rarely were donations sufficient to provide the food, shelter, and other necessities. Even if they came from an upper-class family that looked down on manual labor, their religious vows and the necessities of their lives meant they had to deal with the basics of life every day. They were highly motivated to find ways of increasing the time they could spend on their religious activities, so the innovative ones among them and the villagers and peasants who helped them found ways of improving the waterwheels, mills, tools, and plows. They developed new tools and methods of raising, storing, and preserving food and of constructing buildings and pens. Food and manufactured items were sold to bring in revenue.

In such an environment there was less fear of offending someone in power, so these kinds of innovation were encouraged. Since the monasteries were made up of religious people, no religious taboos were being violated. Once inventions were shown to work on the monastic estates, they were put to work on the feudal estates as well. So at the monastic estates the fears generated by the autocracy, religious authorities, and class structure were much weaker, allowing blue-collar innovation to flourish; in addition, some of the benefits could be passed on to ordinary people.

In the Middle Ages the class structure in Europe gradually opened up. Increased trade with the Middle and Far East increasingly included merchants who weren't born into the aristocracy, at first in coastal cities like Venice and Genoa, then in the coastal cities of France, Holland, Spain, and England. Artisans banded together to form guilds. The kings had to share power with them, and the people consequently became less fearful of the monarchies. The landowning barons fought to diminish the powers of the monarchies, leading to the Magna Carta in England. The American experiment in democracy led to a proliferation of democratic attitudes and institutions all over the world.

It was the absence of fear that enabled the American Founding Fathers to write the Declaration of Independence, which contains these remarkable, eloquent words:

> We hold these truths to be self-evident, that all men are created equal, that they are endowed by their Creator with certain inalienable Rights, that among these are Life, Liberty and the pursuit of Happiness—That to secure these rights, Governments are instituted among

> Men, deriving their just powers from the consent of the governed.

These words express the very opposite attitude from that of autocracies. Autocrats and most monarchs of the times considered that they derived their powers from gods or some higher power, not from the consent of the governed. The governed were the subjects of the rulers and were expected to do what they were told. The notion that ordinary people had inalienable rights would have been considered ridiculous. Even the common people of the autocracies would have had a hard time accepting the concept because they had been brainwashed from birth to believe they had little power over their lives. The words of the Declaration of Independence are as empowering today as the day they were written. Dictators today work very hard to give the appearance that the governed approve of them, so much so that they do everything they can to eliminate opposition; they then boast of getting 90 percent or more of the vote, in spite of how ludicrous their claim appears to the outside world.

All this doesn't mean that people in democracies don't experience any inhibiting fears. Fear of ostracism for thinking outside certain limits is still present in some towns, communities, and churches. Fear of bosses still prevents people from pushing for changes in the workplace. The idea of class is still strongly present everywhere, and people who identify themselves as lower class feel (often correctly) that if they try to do anything significant, someone in the power structure will go after them. Others fear the humiliation that goes with failure. Others feel constrained by their lack of education. Whatever it is that makes people fearful, it has an inhibiting effect. One could argue that these fears are self-imposed at the individual level, and thus it is not the concern of the society at large. While such an argument has some merit, the society does exert an influence through its institutions, which can diminish or perpetuate such fears. Teachers and religious leaders should make clear that questioning current ideas (even educational and religious ones) and current methods of doing things is a positive endeavor, one that can benefit everyone. What we do know is that as the sources of fears diminish, more people feel freer to innovate.

The historical record is clear: in innovative societies citizens are free from fears that prevent them from making changes to the status quo. They do not feel inhibited by their family background, their position in society, or their lack of wealth from trying to improve their situations or attempting something new. They do not feel that the authorities will view their efforts with suspicion. They feel free to question how things are done and to suggest new ways of doing things. They feel free to experiment, to risk failure.

It is obvious, then, that a government that wants to encourage innovation among its citizens should take steps to reduce the fears described above. This means not only making sure the laws and practices of the government itself don't inhibit citizens from advocating changes, but also preventing other centers of power within the society (i.e., the military, police, corporations, and other private institutions) from perpetrating these fears. Governments and their laws should be designed to ensure that the widest number of people possible, including women, minorities, and immigrants, feel free from fear.

11. Access to Information

Intercultural contacts; seeing is understanding, sometimes; the role of transportation and trade; secrecy doesn't help; the printing press changes everything; libraries and journals; accessing information in the digital age

An innovative society will make it possible for its citizens to have access to the kinds of information that support innovative activity, and it will provide the educational opportunities necessary to understand the information. Access to information refers not only to the means and convenience of access to information but also to the relevance of the information and, even more important, to the number of people within the society who have this access. If only a select group like a priesthood or emperor's court has such access, the creative energies of the rest of the population will not be harnessed, to the detriment of all.

One of the most important sources of access to information has historically come from intercultural contacts, as mentioned in chapter 6. Ports and well-established trade routes provided the means whereby the ideas of different societies and cultures could be exchanged, and this spurred innovative activity. As a consequence, cities near seas and oceans became important conduits of new ideas.

The earliest access was by direct visual and manual inspection. Many practical inventions are relatively easy for people to understand and imitate once they see them. At a museum, for example, it's often possible to figure out how ancient devices worked or why they were constructed in a certain way simply by examining them as they sit on the shelves. The motivation for making improvements in knives, swords, and other weapons is usually quite evident. Levers, simple pulleys, cranks, grinding mills, lifting devices, catapults, ladders, wheeled carts, and hundreds of hand tools can be similarly analyzed. Even today, patents are awarded for devices whose method of operation is apparent from the drawings.

Other devices are not so obvious, but they can be analyzed by workers with special skills. While most people might not recognize the novelty in a new loom design, for example, a person who builds looms will recognize it immediately. A new method for cutting or shaping granite can be recognized by a stonecutter, and, once seen, the method can be copied. Some improvements can be deduced from an accurate drawing or verbal description, while others may require actually seeing the device in action.

Other inventions involve improvements that aren't so obvious, even to trained people. Many such improvements require the use of special materials with properties that aren't apparent from a visual examination. Someone attempting to copy a catapult, for example, might not realize that only certain woods cut in a certain way would ensure that the catapult would operate without breaking. Knives might look identical, but the properties of the metals and thus the utility of the knives are quite different. If the success of an invention depends on the properties of the materials used, then a mere visual observation of an invention in use may not be enough to enable its duplication. It will require special knowledge of the materials used and of the manner in which the materials were formed. Such special knowledge was often deliberately kept secret so the maker could maintain a hold on the market and keep doing his job. Sword makers who developed their own methods of hardening and shaping steel guarded these methods carefully.

The need to protect such secrets is at odds with the need for access to information, and this tension between conflicting needs has been, and continues to be, a major issue. Under autocracies, the need for secrecy dominated all other considerations. Military weapon designs were given special protection for the obvious reason that it gave the autocrat an advantage over his enemies. Even the techniques and materials involved with royal or religious buildings, decorations, and jewelry were held close, so as to make the finished products unique and special.

All such knowledge can be kept secret only for a limited amount of time. Eventually the techniques become known within the profession, much like patented techniques are today. Once that happens, the craftsmen will be taught the techniques as part of their training, and the innovators will be off making other improvements, which eventually will make their way into the common knowledge base.

Fortunately for progress today, it is usually in the interests of industry and commerce to patent such ideas in order to protect them, at least for long enough to develop and sell products. Patents are published, making them immediately accessible to the general public. The inventions and the thinking behind them can then be used as starting points for further innovation. This is an important reason why the patent is considered to be a key development in modern inventive activity.

Today the access to technical knowledge is amazingly wide, and the internet has hugely amplified this accessibility. In the past, however, there were a number of barriers and limitations on the access by the people to knowledge essential to innovation.

One such barrier has historically been geographic. For a long time the ability of people to get from their homes to the sources of knowledge was very limited. When the sources of knowledge were the mines, furnaces, or machines themselves, very few people were able to conveniently travel to see them, even if they were allowed to do so by the local authorities. The trade that took place between the various societies carried the final products that used the new techniques. Trade, of course, involved materials, products, and animals that could be sold or bartered and that were considered desirable by foreigners. Different kingdoms had different metals and other materials, different plants and animals, and different skills, so it makes sense that trade has been around ever since there were tribes.

Many of the products of trade aroused the curiosity of rulers who were motivated to find out more about exotic new materials or techniques with an eye toward duplicating them at home. For example, new hard materials would be seen as potentially useful in developing weapons for the military. If the local experts couldn't duplicate the products by inspecting and analyzing them, they would try to get to their source to learn more about their production, and spies were frequently used. If this didn't work, the ruler always had the option of capturing the means of production by force.

White-collar knowledge was kept in the form of scrolls and books, at first in temples and in the autocrats' buildings, and later in libraries. These were located primarily in cities, so access was limited to those who lived nearby and were allowed to read them. This was the genesis of the conventional wisdom that innovation occurs primarily in cities. With our white-collar/blue-collar distinction, it is evident that this notion is not valid for blue-collar innovations such as those in agriculture, mining, and fishing. For innovations in the trades and commerce, however, the cities offered frequent opportunities for contacts and the exchange of ideas. Instead of access to information being deliberately provided by the rulers, the access was provided simply by the interactions of human beings with each other.

Intercity trade provided access to information for the commercial participants. As trade and ship traffic increased around the Mediterranean, the increased contact between the cultures resulted in an exchange of ideas that spurred both kinds of innovation. People traveled between cities, learned the local languages, and gained access to foreign scholars and their writings. In this manner Greek mathematicians like Pythagoras were able to build on the geometry they had learned in Egypt. Their students then passed on the knowledge to people in other cities, and important ideas got propagated. Alexandria, for example, was a center of innovative activity that attracted creative people from all over the Mediterranean, such as Archimedes from Syracuse.

What we today call postal services have a history that goes back to ancient China, where messages were sent between government officials by horseback. Since then, many civilizations, notably the Roman Empire, have done the same. Merchants also carried messages, some no doubt containing orders for future deliveries, by land and sea. European universities organized students as couriers to carry documents in the Middle Ages.[1] Around 1400, postal services that carried documents between people were established in England and Italy, and these soon spread to other regions. Postal service by land and sea helped speed communications and carry documents everywhere. Then an explosion of communication and transportation technologies occurred in the 1800s: faster ships, trains, the telegraph, underwater cables, telephones, and fax machines. We are now very fortunate to have satellite communications and the incredible internet. Today the geographic barriers have been virtually eliminated for a large percentage of the world's people. Knowledge of advances in practical and theoretical matters now propagates rapidly around the world.

However, there have been other barriers to access to information besides geographic ones, and many of these are still problematic. Language was a significant barrier, and remains so; the huge number of different languages and local variations meant that information about innovations didn't cross language boundaries very rapidly. While merchants who traveled a lot learned enough foreign words to conduct business, only a few of them learned enough to understand technical subtleties. Those who did were proficient only in a few languages, so their direct knowledge was limited to a few geographic areas. So ideas and accurate descriptions of inventions were propagated quite slowly between societies.

Another important barrier, especially in blue-collar innovation, was the economic motivation to keep novel techniques secret. Before patent law was developed and enforced, once the inventor's methods were made public, he often could no longer earn a significant income from his efforts—others would immediately utilize his techniques. As a consequence, artisans and guild members took great pains to deny access to the crucial information. This made perfect sense to them, but it did significantly slow the spread of many ideas.

One positive feature of the large autocratic societies was that in conquering other territories, the conquerors' languages were learned over a wider geographic area, which reduced the language barrier. This was true for empires of all kinds. The Greek language was used all around the Mediterranean for several hundred years. After that, the Latin language of the Roman Empire made it possible for scholars and merchants to communicate across Europe throughout the Middle Ages and beyond. The courts of the Chinese emperors enforced a top-down version of the written Chinese language throughout the empire, and this contributed to increased literacy there.

Recording ideas and events was recognized by the autocrats as a valuable means of expressing their power and legitimizing their rule. Scrolls of papyrus were used to record transactions, praises to the reigning kings, historical accounts, and philosophical and religious ideas. After Egypt prohibited the foreign sale of papyrus, parchment (made of animal skin) was used in scrolls. The profession of the scribe became important. In the early days of Christianity, the book as we know it was invented. The books had covers made of wood or leather and consisted of pages rather than continuous rolls. Titles were written on the covers for ready identification, and the stiffness of the books made it possible to conveniently place them on shelves. The early monasteries became major purchasers of new books. Paper, a wood-based product that could be manufactured much more cheaply than parchment or cloth, replaced papyrus and parchment over a period of a few centuries for use in books. These advances set the stage for the huge libraries of the Islamic Empire, including those of medieval Spain.[2]

The portability of books coupled with the armies of scribes and scholars who translated the collective knowledge of the world into Arabic significantly reduced the geographic barrier. Scholars came from all over Europe to Spain to learn Arabic, to glean knowledge from the books, and to translate this knowledge into Latin, from whence it could be read by other European scholars. These scholars would then study the books over a period of time in their own homes and schools and pass them on to students or to university libraries.[3]

The invention of the printing press in about 1440 accelerated this process. As presses proliferated, white-collar knowledge began to be exchanged between literate people. This happened at a time when even much of the nobility was illiterate, but everyone who could afford it wanted a Gutenberg Bible, and literacy spread quickly throughout the nobility. Soon almost everyone wanted to be able to read, although the goal was not realistic for most poor people, which meant most people. But the word was out, and those who could do so tried to become literate. *De Re Metallica* (1556), described in chapter 5, was one of the earliest blue-collar books to be printed. With many illustrative plates, it described in great detail machines for mining and iron-working, furnace design, metallurgical techniques, techniques for diverting streams into mines and troughs, and waterwheels. It also detailed a wide variety of machines involving ropes, pulleys, and gears for lifting, pumping bellows, and stamping metal. Even then the saying was known that a picture is worth a thousand words. The author Georgius Agricola, born Georg Bauer, provided enough detail in the plates that even an illiterate worker could learn much about metal working from studying them, with some help from a literate crew leader. From this time on, blue-collar innovations would become increasingly communicated through the written word. The potential access to information had suddenly taken a giant leap forward for all kinds of innovation.

Libraries have historically been the most important repositories for white-collar knowledge, so their availability has determined the access to that information. The ancient libraries contained clay tables, papyrus scrolls, and parchment documents. They were accessible only to a privileged few—namely, priests, scribes, and scholars of the courts. Later libraries, like the famous one at Alexandria, were available to scholars from around the Mediterranean, and Roman libraries were available to citizens of Rome who could afford the privilege. In some circles libraries became status symbols, and private libraries proliferated. In early medieval times monasteries established libraries that, while primarily containing religious texts, contained other books as well. Islamic Houses of Wisdom in the tenth century were probably the first to be open to the literate public, and in the seventeenth century public libraries were established in Europe and the United States. Prior to the eighteenth century, most libraries had to be accessed on-site (i.e., books had to be read and copied within the walls of the buildings). In the 1800s, when there was a dramatic growth in literacy, public libraries were established and funded by local governments and offered lending programs that let people take books home to read. By then the libraries were the repositories of blue-collar knowledge as well.

Historically, innovative societies have provided access to the collective knowledge for a select group of people. In early societies, this meant the priesthood, some builders, the military, and the royal court. As technologies and writing expanded, specialties developed in architecture, construction, mining, stonework, metalworking, and writing materials, and these were propagated both orally and through pictures and writings. Some rulers recognized the value of storing this collective wisdom, and they established libraries to store it. Much blue-collar knowledge was no doubt lost because of the inadequacy of oral communications to accurately transmit detailed and subtle information and because little was actually written down. The inventions that made the publication of written information inexpensive set the stage for a huge expansion of the access of knowledge to the majority of the people, although that was not to be realized until literacy became widespread within the last two centuries. The inventions of the 1700s were made possible by, among other causes, the loosening of the restrictions on access to information and the modest early increases in literacy.

In the last two hundred years, there has been an explosion of information, and most of it has been accessible to a large fraction of the population. Public libraries have proliferated, as have books and scholarly and professional journals on every conceivable subject, as well as newspapers and magazines that have trumpeted the innovations of the world. Most recently, the internet has made a wealth of information available to anyone with access to a computer and an internet connection.

Today most societies provide their citizens with a level of access to knowledge that would have seemed like science fiction a few centuries ago. Access to information has practically become a universal right.

12. Guidance in the Use of Information

Learning from the masters; learning from peers; peer guidance and the printing press; putting the knowledge of experts in perspective; institutions attempt to guide invention

The third essential condition for innovation in a society is how well it provides guidance to its young people. If there is a large body of relevant and useful knowledge potentially available, but the resources for education are not properly supported and prepared, the young citizens will not be adequately prepared to use the knowledge to sustain and improve on the capabilities already developed in the society.

In early societies the tribal elders, both men and women, spent a great deal of their time passing down the essential history of the tribe to the younger people, including utilitarian information about food, health, and housing essential for survival. Since the tribe was illiterate, this information was passed down by repetitive demonstrations of things like stone-shaping, cooking, and tent building, by encouraging active participation in hunting and food gathering, and by teaching the identification and processing of medicinal plants, barks, and leaves. The information was reinforced by chants, songs, poetry, drawings, and stories depicting the tribal lore and essential knowledge. There are still tribes today where these are the ways practical information is propagated from generation to generation. Of course, as tribes fought, this knowledge was sometimes destroyed, while in other instances knowledge got added to that of the conquering tribe.

When civilizations developed, the family and village mentoring continued to be the way specialized blue-collar knowledge was passed down among the working people. A different mode developed in the courts of the autocrats, where white-collar knowledge was born and grew. There, people developed bodies of knowledge in astronomy, medicine, horticulture, weapons, and construction. Unlike today, these areas of knowledge were much less differentiated then, and the same people were often knowledgeable in more than one area. The astronomers and religious elite were able to operate independently of the common people, but other administrative specialists had to work with the commoners. The court physicians had to get their herbs and chemicals from farmers and workers; they obtained their knowledge from the collective wisdom of the common people. The horticulturists who designed and supervised the royal gardens likewise had to work with farmers to obtain plants and learn about the intricacies of planting, pollinating, nurturing, and maintaining them. The military leaders had to deal with workers who chopped trees, shaped wood, mined metals, built carriages, trained animals, made clothes and shields for warriors, and formed swords, bows, and arrows. The architects and builders had to learn the techniques of locating quarries, removing marble and granite from the ground, stonecutting, and transporting heavy objects over long distances; these they could not learn from theorizing in their offices, but rather they had to learn from the practical knowledge of the men doing the work in the mines and fields. Of course, the leaders and their lieutenants learned what they needed to know and had a strong interest in teaching this knowledge to their eventual replacements. Between these elite leaders and the workers, the knowledge was passed on by the elders guiding the younger ones.

Once writing was invented, the potential was there for recording the techniques and methods used. I haven't come across any evidence demonstrating that scrolls and tablets were used to record such techniques prior to the Greeks and Romans, although they well may have; probably, the techniques were simple enough that oral communication of this knowledge was considered adequate. In the sixteenth century, Agricola[1] mentions that no part of the subject of mining is "fully dealt with by the Greek and Latin authors whose works survive," which indicates that these early authors did indeed use the written word for preparing what we might today call instruction manuals. Given the limited level of literacy at that time, the manuals would have been read and utilized by technically oriented offspring of the nobility. They would then have trained the less literate workers.

The system of guilds in medieval Europe served to effectively propagate knowledge down through the generations. The members of merchant and craft guilds trained not only their own sons but also the sons of other guild members. This guidance involved hands-on training and an organized, progressive hierarchy from apprentice to journeyman to master. One of the reasons for their collective action and cohesiveness was that the guilds formed a protective shield against over taxation and confiscation by predatory rulers.

At about the time of the invention of the printing press in the fifteenth century, new and sophisticated inks, metals, and glass were being developed, and the fabrication techniques were becoming more complex and difficult to accurately depict orally—so were the techniques for mining metals. So it was fortuitous that the capability of printing relatively inexpensive books was developed in time to enable people to describe in words and illustrate in pictures the complex techniques that were coming increasingly into use. The printed word gradually became increasingly important in providing guidance in the use of information needed for all kinds of blue-collar innovative activity.

Nowhere was this more evident than in the eighteenth century in Britain, where mechanics, artisans, and millwrights found it exhilarating and economically beneficial to learn to read and equally important "to measure and calculate." [2] They met in pubs and coffeehouses to discuss new ideas and to learn reading and practical mathematics from each other and from scientific experiments that were taking place at the time. Some lectured around the countryside and wrote books and articles. "'Useful knowledge' … became the buzzword of the eighteenth century." This was peer guidance at its best.

In the more abstract areas of learning, the written word made it possible to record mathematical discoveries, astronomical observations, and surveying techniques that successive scholars could build on. The Greeks developed the concept of a public building where people could hold forth their thoughts on philosophy and politics, where they could teach and be questioned by citizens. Aristotle took this one step further and founded a formal school called the Lyceum. Similar schools were established around the Mediterranean, the most famous being the one in Alexandria, where the well-known library was founded. It appears to have been the first research institute, a place where some scholars were paid to think and experiment, as well as a place for other scholars to come to exchange ideas. This was an early example of guidance in the use of information by peers. Later, Arabic and Islamic Houses of Wisdom were established in the Middle East, the Mediterranean, and in medieval Spain. Soon thereafter universities were established in Europe: the University of Bologna in 1088, the University of Paris in 1160, and Oxford in about 1167. At these and many other centers of learning around the world, scholars taught students and each other what they knew about a variety of subjects, including astronomy, mathematics, and botany, as well as philosophy, history, medicine, and religion (to which we probably ought to add the beliefs and assertions about alchemy and astrology). They wrote treatises, visited each other, and exchanged ideas.

While institutions of learning were taking hold, the world of commerce was growing. The learning of new languages, the methods of doing business with foreign merchants, and the specialized knowledge of navigation and the shipbuilding business required new professions. Captaining a ship required entrepreneurial skills to convince financiers and merchants to support their missions, language skills to deal with people in foreign ports, navigation and sailing skills to get from place to place, and managerial skills to train and direct the sailors. These skills and the knowledge of the operation of practical inventions got passed on by direct teaching and experience through a hierarchy of captains, mates, and other ships' officers. In the city-states of what is now Italy, trade grew dramatically, and by the 1300s schools were established that were attended by students other than the sons of the nobility. For example, in Florence almost half of the boys that lived there attended these schools. In addition there were "abacus schools" that trained promising students in bookkeeping and record keeping.[3]

One of the next developments was that of learned societies, which extended the peer guidance concept to include scheduled meetings and printed journals and evolved later into professional organizations. Instead of residing at a physical institution, the location of the societies was more distributed geographically and temporally; individual communications, by mail and through occasional meetings at different locations and for short periods of time, expanded the idea of courses at universities. One of the earliest examples was the Accademia dei Lincei, or the Academy of the Lynxes, which was established in Rome in 1603 and had Galileo as one of its early members. Benjamin Franklin founded the American Philosophical Society in 1743 initially as a discussion group, but it eventually became a prestigious institution with its own journal and prizes for outstanding achievement in navigation, astronomy, and natural philosophy. By 1800 there were many such organizations all over Europe, and today they are all over the world. By identifying like-minded people in the quest for knowledge, one can now get individual or collective guidance in a specialty field. By attending conferences and reading professional journals, one can learn the latest thinking on just about any subject whatsoever.

Even with all these advances, literacy remained limited until public education became widespread. The concept of public education was one important driver of the explosion of inventions of the nineteenth and twentieth centuries. Instead of just 2 or 3 percent of the population being educated, eventually the majority of people in the more fortunate societies were trained to read, write, and, to some extent, think. This caused a many-fold increase in the number of people who could take advantage of the written wisdom. Widespread literacy was one of the strongest factors in the explosion of innovation that took place over the last two hundred years. The advocacy of public education by John Adams, Benjamin Franklin, and other Founding Fathers of the United

States was one of their greatest gifts to posterity. While illiterate people could still invent, they were at a tremendous disadvantage compared to literate people, who could much more quickly learn the ideas, tools, materials, techniques, and thought processes of inventive people who came before them.

Traditional education focused on the transfer of the knowledge of the society to the students, and teachers assumed quite reasonably that this would prepare students for life and give them a foundation for any further thinking they might do. Unwittingly, it also encouraged an attitude of excessive reverence for past thinkers and past knowledge, and to some extent discouraged original thought. This did not stop some fearless pioneers, however, and it gradually became more acceptable to question the old answers and to branch out into new topic areas such as machines, animal husbandry, mining techniques, economics, medicines, and many others. These developments led to the establishment of specialized technical and engineering schools, and eventually colleges and university departments, devoted to these subjects. Today it is possible to obtain specialized knowledge and guidance in a wide variety of subjects.

Mentoring by peers took a major step in the late 1800s with the development of the industrial laboratory, pioneered by Thomas Edison's Menlo Park laboratory in New Jersey. By working together under one roof, as in ancient Alexandria, researchers could bounce ideas off each other, learn specialized knowledge rapidly, and compete with each other in the development of new ideas. Industrial research centers, both as separate bodies and as auxiliary departments in large companies, sprang up in Germany and the United States, as well as throughout Europe. Governments got into the act by establishing research centers both as separate entities and at universities. Today such institutions number in the thousands. While they have experienced varying degrees of success, many of the most important inventions that affect our lives today have their roots in the research conducted at these institutions.

It is clear that societies that educate their people will be more innovative than those who don't—no surprise there. But to be effective, societal guidance should go beyond the mere teaching of reading and writing, although for the most innovative individuals that may be enough. It is essential to teach problem solving in all subjects—it's not enough to impart yesterday's knowledge. One serious trap that many educators and teachers have fallen into over the years is to believe that their only goal is the teaching of knowledge. While knowledge is essential, it is not enough. First of all, the knowledge taught will rarely, if ever, be enough to support innovative activity. Innovators need specialized knowledge that can only be acquired from outside sources or from experimentation. Second, students must be given the encouragement to examine the events and knowledge and try to figure out the assumptions behind them. Once they learn what these assumptions are, they can then critique them and postulate what might be possible under different assumptions. Modern educators recognize this issue and have developed teaching techniques, such as inquiry-based learning, to incorporate inquisitive thinking in the classroom.[4]

It is difficult for dictatorships and autocracies to promote independent thinking because they have such a strong stake in controlling information. As a consequence, their textbooks and reading materials usually praise the status quo, venerate the rulers, and discourage the reading of foreign accounts as being potentially subversive of authority.

While reading about the lives of innovative individuals, it is difficult to avoid thinking that there are probably many high-level achievers who don't need training in how to be innovative—they just need society to provide education in the basic tools and access to information, but otherwise to just get out of their way. While this notion has merit, it is also true that those of us not endowed with that same talent and drive can still make significant contributions to society if we are provided with proper guidance in the use of information.

13. Societal Rewards for Innovation

Losing the zero-sum-game assumption; recognition is nice, but jobs and money are nicer; rock stars or inventors? the role of patents

The fourth essential condition for an innovative society is that it rewards innovative people. These rewards take a variety of forms: positions, jobs, titles, awards, fame, status, respect, prizes, and money. Less innovative societies will have fewer rewards, and they will place more restrictions on who can receive rewards and on the size of the rewards. Conversely, highly innovative societies will have many ways to reward people, will have more reasons to reward them, and will be more generous in the size of the rewards. For example, does an inventor receive a medal, a patent, or a means of making money?

To take advantage of its benefits, there must be a conscious recognition throughout the society that innovation is a positive force that potentially benefits everyone. If it is seen as a zero-sum game, in which if someone benefits (e.g., a carmaker), then someone else loses (e.g., a carriage maker), the result will be a strong resistance to new ideas and inventions. Yet that was the prevailing view of most governments and most people until the last few hundred years.

Indicators of the Value of Innovation to the Cultures

Cultural values determine the societal rewards for innovation. These values may not be hostile to innovation, but they can still discourage it if it is considered unimportant. Indifference can kill the motivation for even considering new techniques and ideas. If young people are taught that coming up with new things is frivolous or nothing but toy-making, then they are not likely to be highly motivated to spend their time in inventive activity. Rewards are a result of general attitudes toward innovation; they reflect both the dominant culture and the subcultures of the society.

One indicator of cultural values is the choice of heroes. Are the prevailing role models primarily military, religious, government, athletic, and entertainment figures, or do they include scholars, scientists, inventors, and entrepreneurs? In ancient times, there were not even any words for scientists, inventors, engineers, or innovators. There were priests, scribes, and court astrologers who made occasional contributions to white-collar knowledge, and there were military leaders of soldiers and managers of slaves who made similar contributions to blue-collar knowledge. No doubt some of them achieved heroic status on a local scale. It is only relatively recently that people like Benjamin Franklin, Thomas Edison, John Harrison, Louis Pasteur, and Albert Einstein achieved star status in the mind of the public.

Another indicator is the kind of occupations that are considered desirable. Do any of them specifically involve inventive activity? Do they involve applying existing knowledge, or are there recognized opportunities within them to make significant changes? It has been only in the last few hundred years that engineering came into being as an occupation. Today there are dozens, even hundreds, of different types of engineering careers that a young person can choose to pursue. In ancient times boys were usually expected to pursue the craft or occupation of their fathers or other older role models, most of whom placed a high value on workmanship rather than novelty; girls were totally ignored.

Another indicator is the attitude of the culture toward respect for tradition. Is tradition valued to the point where criticism of the ancient thinkers is considered disrespectful or even heretical? If so, then people's ideas will be judged on how well they reflect and support the conventional wisdom. In an innovative society the general attitude will be one of welcoming reconsideration of old ideas, with the caveat that one needs to have a convincing case for contradicting the ancients. If you're going to challenge Einstein's theory of relativity, all is well and good, but don't think it's going to be easy. Overemphasis on tradition can retard innovation; too much veneration of ancient wisdom and historical figures can discourage people from questioning previous assumptions. There needs to be a balance between respecting the work and the thinking that went into previous efforts and recognizing that there are usually ways in which they can be improved and even replaced with better ideas or techniques. In societies where tradition was venerated to the point of worship, this attitude was unconsciously extended to all phases of life. Many ancients were highly intelligent, but they all made errors in judgment and were limited by the state of knowledge at the time. Aristotle didn't have the benefit of Galileo's telescope to see the fallacy with thinking that the planets rested on crystal spheres centered at the earth; he also did not have the benefit of accurate measurements that showed that the planetary paths were not circles around the earth—they weren't even exactly circular around the sun. Too much veneration for ancient geniuses led to the assumption that just because they said something, it must be true. As was pointed out earlier, even their students and colleagues knew better. Blue-collar innovation was less affected by this phenomenon, partly because the names of the inventors were not known. The attitude that "I don't know why, but it's always been done this way" or "my father did it this way, and that's good enough for me" was certainly widespread (and still is in many places), but if one actually demonstrated an improved tool, it had a good chance of being accepted because the benefit would be readily evident.

Another indicator is the rigidity of the class structure, in particular its resistance to upward mobility. Societies in which people's opportunities throughout their lifetimes are almost totally determined by their families of birth limit the rewards for innovative activity. The rewards of positions, titles, business monopolies, and recognition by the dominant culture go almost exclusively to the upper class. These classes have typically been a small fraction of the total population, most of whom did not have access to these rewards. The potential innovative energies of the society have thus been suppressed, rather than enabled. Of course, the people of the ruling classes have typically considered the people of the lower classes to be intrinsically inferior—less intelligent, less refined, less capable, and less ambitious. They were there to do the essential grunt work of raising crops, digging up metals and granite, hauling food and materials around, constructing buildings, and waiting on the "important" people. The ordinary people no doubt considered themselves to be important to each other, but generally they bought into the concept that the ruling classes were superior. Up through the 1600s, this attitude was prevalent throughout the world, but the forces that would change it were gathering steam and would eventually change the world.

Religious rigidity is another indicator. While religion can encourage innovation, we have seen how religious authorities can spread fear and can suppress ideas that are judged to be counter to the existing dogma, especially white-collar innovation. We also saw that religious communities can develop a suspicion of people who study certain subjects. While the fear involved is not immobilizing, it can still have a chilling effect on generating new ideas. Even when fear is not involved, religions can discourage innovative thinking of all kinds if they promote a fatalistic attitude in people. If people are taught that the world is static and unchanging, that the recorded wisdom of the ancients is not subject to scrutiny or criticism, that acceptance of one's fate is a virtue, then there will be a subtle pressure for people to accept things the way they are and not look to make improvements. Such an attitude affects both blue- and white-collar activity. If the dominant culture adopts this attitude, it will be unlikely to provide rewards for any activities that promote change.

A negative attitude of the culture toward change can discourage innovation. It may be due to features of the dominant religion, as noted above, or it can be inculcated by other sources. Any innovation or invention is going to change how something is done, and someone is always quite happy with the status quo and doesn't want it changed. We see this every day in our work, our organizations, and even our families. Often insecure people in power have found it in their interest to promote resistance to change through the dominant religion and government institutions. For innovation to thrive there must be counterbalancing attitudes that welcome, or at least tolerate, new ideas.

Complacency is another indicator. Empires that were successful in the traditional sense were particularly susceptible to this disease because they attributed their previous successes to their cultural superiority. This attitude reflects a widespread fallacy of human beings to believe that the culture they live in is the best, and as such it deserves to, and will, last forever. In fact, however, history shows that our lifetimes span a relatively tiny period, and it only takes a century or two for empires to be built or disappear. Finding any society that lasted for a thousand years in more or less the same form is rare, indeed. China, in particular during the Ming Dynasty (1368–1644), considered itself to be the center of civilization; even though China was in a position to establish profitable commercial trade throughout the Middle East, India, Africa, and even Europe, it declined to do so, taking the attitude that "they have nothing we would want." In so doing, the Chinese denied themselves the opportunity to take advantage of innovations from other countries. The Islamic Empire, after becoming the beacon of enlightenment in the Mediterranean, lost its innovative ways; its leaders interpreted their successes to conclude that God favored them, so there was no need to continue to explore new ideas. The Ottoman Empire looked at the mechanical clocks, cannons, and other inventions of the Europeans and dismissed them as toys; they made interesting gifts, but had no real value otherwise. Later, the Spanish and Portuguese empires, rich with the booty they had captured from the New World, made no serious attempts to set up sustainable economic systems, believing they did not need to manufacture things themselves because they could buy anything they needed. In all these cases, complacency was a large factor in the fall of the empires. The previous success encouraged the fantasy that the empire was naturally superior and that the good times would continue indefinitely. Potentially innovative activity ran into the attitude of "why change something that is working well?" Complacent societies will subtly (or not so subtly) discourage innovative activity.

The indicators described above, however, are what the term implies: indicative. They are broad and unspecific; singly or collectively, negative indicators don't cause a total suppression of innovative activity. There were a few inventions in these empires in spite of the lack of encouragement. Some individuals were in situations that insulated them from the negative societal attitudes, and they received enough recognition or other rewards to motivate them. But the overall effect was to prevent societies from reaching the tipping point of recognizing the paradigm that technology begets technology[1] or identifying the invention of invention.[2] Such concepts did not take root in societies where these indicators were generally unfriendly to innovative activity.

White-Collar and Blue-Collar Rewards

White-collar innovation in particular has been affected by the attitude of the dominant culture. The pharaohs and kings of old decided which ideas would be deemed acceptable and what topics could be explored. The Greek culture was different, and it allowed philosophers more freedom to expound on their theories of humanity and the universe. The medieval Catholic Church recognized the yearnings of smart people to exercise their minds and set up institutions like the Jesuit order so they could do so, within limits. The Venetian Republic and other city-states, which were run by groups of rich families, set up universities and sponsored philosophers, artists, scientists, and other thinkers. The dominant cultures encouraged creative thinking and expanded the scope of the subjects they could explore. While some religious subcultures within were hostile to these activities, the dominant culture prevailed.

In the early world of white-collar innovation, a position at court was considered the ultimate reward. Later, rich noblemen and royalty sponsored esteemed intellectuals, and universities offered what we today call professorships. The Church assigned promising intellectuals to Jesuit institutions that had libraries and an atmosphere of inquisitiveness. The inventor of algebra, Al-Khwārizmī, was a scholar at the House of Wisdom in Baghdad, an institute established by the Abbasid caliphate. Galileo obtained a position at the university in Padua and later was sponsored by a member of the Medici family. While Albert Einstein was a clerk at the Swiss patent office, he wrote his seminal paper on special relativity and was rewarded with a lecturer position at the University of Bern. The great Library of Alexandria was one of the first research institutes, and it provided stipends and housing for some fortunate scholars and their families. It afforded recipients an opportunity to devote their energies and time to innovative activities, and in return they were expected to bring honor to the institution. Institutions like this were rare until the last few hundred years.

Having positions involving an income was quite important, especially if the recipients were not independently wealthy. Wealthy thinkers had the luxury of pursuing their interests without being distracted by other demands on their time. A number of scientists were independently wealthy, including Émilie du Châtelet (1706–49),[3,4] Robert Boyle,[5] and Otto Gericke.[6]

The most prestigious and tangible type of societal reward is formal recognition by the ruling elite in the form of a title or an award. The famous scientist Isaac Newton is often today called *Sir* Isaac Newton in reference to his knighting by Queen Anne in 1705. The British tradition of knighthoods is probably the best known of the formal honors, but many countries have their own versions. One measure of the level of societal rewards is the number that are received by innovative individuals. Today one of the ultimate honors is the Nobel Prize, awarded to innovators and other creative people throughout the world.

Blue-collar innovation prior to the last few hundred years was more affected by the subcultures rather than the dominant culture because the big rewards of major positions, titles, and money were not available to people in the lower classes. This was due in large part to the arrogance of the elite classes, who avoided and demeaned manual activity and dismissed the efforts of the lower classes as trivial. This was probably not a universal attitude because, of necessity, some elites had to act as managers of the farmers, landscapers, craftsmen, builders, soldiers, and workers. It's difficult to believe that none of these managers had any grudging respect for the ideas of the common people, but if they had proposed some significant rewards to inventive workers, they probably wouldn't have gotten much support from their peers. It's more likely that the leaders took the credit for any improvements. Inventors in the lower classes would have been motivated by the approval of their neighbors, competition with other craftsmen, or possibly a promotion to a good position by the manager of the farm, mine, or quarry. That is, status, fame, and respect motivated most inventors, while a few could hope for recognition that came with tangible benefits.

Being admired by your neighbors is a powerful incentive, more so when other rewards are limited. This was sufficient motivation for most of the blue-collar innovations before 1700 throughout the world. We shall see in later chapters that as widespread recognition became possible for inventors and other creative people not born into wealth, as the potential for making money became more real, and as literacy spread, the blue-collar/white-collar distinction became less pronounced, and societal rewards of all kinds accrued to all sectors of society.

Societal Rewards throughout History

As I have emphasized, most societies have encouraged some kinds of innovation and have rewarded efforts at least in circumscribed areas. The early autocrats rewarded their chief architects and builders with high positions and money, and these, in turn, rewarded individuals who came up with new ways of creating, forming, or decorating beautiful objects. Military leaders rewarded men who designed and built new weapons, a typical reward being money, an award, or some position of status. Typical military organizations had hierarchies whereby the elite had the top command and strategic positions but where lower ranks of officers involved considerable direct contact with the soldiers, and these were available to promising commoners.

In ancient times the autocracies had courts that served the administrative needs. There were priesthoods of the official religion, as well, often also under the control of the autocrat. These people formed the educated classes, so any white-collar innovation would come from them—in particular, those pursuing astrological, astronomical, or architectural knowledge. Blue-collar innovation was stimulated by the need for the construction of buildings and the crafting of their adornments. For example, new ideas for finding, shaping, and cutting new materials were encouraged, but the elite scorned this kind of work. Outside these circumscribed areas, however, innovative activity could readily run afoul of the priests or the ruling classes. This is the classic dilemma of autocracies, where loyalty trumps progress.

This brittleness began to crack with the Greek city-states, which developed the concept of the master philosopher as a nonreligious, nongovernmental position with social status. The Greek society publicly handed out prizes and awards to outstanding athletes and playwrights. Prior to this, rewards given to people outside the ruling circle, and rewards for activities that benefited ordinary people, were rare.

With the breakup of the Roman Empire, the decentralization of power in Europe, the rise of the Catholic Church, and the emergence of a merchant class made it impossible for the monarchies to reestablish autocracies in the classic model. In spite of their best efforts, the ruling classes were unable to inculcate the same fear in the lower classes and found themselves in a more dependent position. In early medieval times, serfs began to leave the farms for the towns that were forming, craftsmen became specialized, and knights became mercenaries and gave their allegiance to whichever monarch or feudal lords could afford them. Peasants began to do piecework for merchants, craftsmen formed guilds, and knights banded together and formed organizations like the Knights Templar. In other words, small power centers developed and acted as a counterbalance to the ruling classes. The people engaged in these groups began to act in their own interest, which resulted in benefits to the societies that even the ruling classes recognized and embraced. Populations grew, and with them came new opportunities for both types of innovations.

By the 1400s enlightened noblemen like Lorenzo di Medici were sponsoring the arts and sciences, and aspiring artists and inventors like Gutenberg, Leonardo, and Michelangelo could rise from the ranks and aspire to public recognition and paid positions. Universities were established that provided paid positions to scholars, primarily for the teaching of existing knowledge. Merchants set up fairs and markets around Europe where goods could be displayed, bought, and sold; the developers of novel products could get monetary rewards for their efforts.

By the 1600s Galileo and other thinkers could get positions as university instructors. During that time loose associations of scholars and scientists developed throughout Europe, forming a new kind of subculture. They formed honorary organizations, like the Academy of the Lynxes, that bestowed respect and status on the participants. Later they formed academic and scientific organizations that enabled communication channels and provided forums for the sharing of information and the recognition of outstanding work.

By the 1700s, governments were actively promoting innovations with monetary rewards. In 1714 the British government offered £20,000 (equivalent to millions of dollars today) to the inventor who could demonstrate a technique for measuring longitude at sea to half a degree.[7] In 1810 Napoleon offered a million francs to the inventor who could demonstrate a practical linen-spinning machine; chapter 15 describes the life of Philippe de Girard, who responded to this challenge. By this time European scientists and scholars were achieving star status in the eyes of the public, beyond the control of any government or religious institution.

More important than prizes and positions of employment was the emergence of the patent, whereby an inventor was granted monopoly rights for a limited period of time to develop his invention into a commercial, moneymaking product or to sell his rights to someone in a better position to do this. Edward Coke, the attorney general of England, was the person who introduced the statute on monopolies in 1623,[8] which he hoped would eliminate permanent monopolies granted by the monarchs and further the interests of artisans and craftsmen by allowing patents to be awarded to the "first and true inventor." Prior to that, patents had been used as monopoly grants by the monarchs to reward favorites at court. The impact of this statute was not instantaneous, but it eventually had a huge impact. In the next century and a half, the patent was mimicked around Europe and North America as its usefulness became more widely recognized.

Modern societies have greatly added to the number and scope of rewards to innovators: patents, royalties, entrepreneurial opportunities, awards like the Nobel Prize, jobs in business, bonuses, and positions in universities; many other honors are given out to innovators from every walk of life. By now, the recognition of the benefits of innovation is almost universal.

Today innovative activity takes many different forms, but the type that has the most economic impact is product innovation, propelled by companies and motivated by profits. The media is continually talking about the "killer app" and the next version of the smartphone or a smaller computer. These products depend on a huge secondary technology base of circuit boards, computer chips, integrated circuits, displays, operating systems, touch-screen controls, data-storage techniques, input/output devices, and data-processing techniques, to name a few, all of which are undergoing significant innovations. These in turn utilize and motivate tertiary innovations in materials, fabrication techniques, and data-handling concepts. Vehicles, appliances, offices, and factories are all being examined to see how to best utilize the new technologies. The monetary rewards for making and selling novel products are probably the strongest motivators in modern societies for innovative activity.

14. Multiplicity of Opportunities

Choices, choices, choices; China doesn't discover the New World; competition creates more choices; the gradual growth of opportunities

Multiplicity of opportunities is a factor largely ignored by historians, but it is one of the most important conditions for encouraging innovation. While it is not strictly a necessary condition for some kinds of innovation, it is essential for sustained innovation. A modern innovative society will provide a multiplicity of opportunities in a number of areas, including those addressed in the previous four chapters:

- *Multiple sources of relevant information.* If a government limits the sources of information to some approved list, it creates a false unanimity of opinion and a false consistency of information, both of which severely hamper innovation. With the wide access to the internet today, this is much less of a problem.

- *Multiple sources of relevant training.* Citizens who can obtain education and training from a variety of schools and universities bring to the society more complete knowledge and competing ideas.

- *Multiple opportunities for rewards for innovative accomplishments.* The more rewards there are for innovative activity, the better. These rewards include jobs in industry and government, the granting of research contracts, and monetary or public-recognition awards.

- *Multiple channels for the dissemination of new ideas.* Having a multiplicity of channels for the dissemination of ideas increases the probability that innovative people can get credit for their efforts. These channels include public patents, trade magazines and other publications, books, and, of course, the internet.

- *Multiple sources of support and funding for the development of new products, techniques, and ideas.* For inventors of products and their entrepreneurial supporters, it is essential to be able to obtain funding from a variety of sources, such as companies, banks, private investors, and venture capitalists.

Governments that are obsessed about maintaining control of the society will, whether deliberately or not, limit these options. New products and ideas always introduce problems to members of the government bureaucracy who don't want to have to deal with very many of them. The primary interest of the bureaucrats will often be to make sure that new ideas don't interfere with existing institutions and monopolies. Providing the opportunities necessary to support innovation involves a certain loss of control, and some societies are not up to doing that.

Another important feature of the multiplicity of opportunities is that it reduces fear. With multiplicity come different factions and power centers that protect their members. Inventors with new ideas that might upset one large corporation will find competing companies willing to use their ideas and protect them from punitive actions.

An Example of the Profound Effects of Multiplicity of Opportunities for Financial Support

A story that dramatically illustrates the power of the multiplicity of opportunities available for financial support involves the contrasting experiences of Zheng-He, a talented Chinese admiral in the early 1400s, and Christopher Columbus, the famous European explorer later in the century. Diamond[1] and Landes[2] were struck by the vivid differences in the outcome of their endeavors, and attributed them to the contrasting opportunities afforded to Zheng-He and Columbus.

Zheng-He was a Muslim adviser to the Ming dynasty who became an admiral in charge of a huge fleet of ships. In 1405 the fleet consisted of 317 vessels and carried 28,000 crewmen and soldiers. In the next three years, a massive effort of shipbuilding and retrofitting took place, culminating in a fleet of over 1600 ships. Some of these ships were massive, some four hundred feet long, with the largest ships designed for royal use. All three of Columbus's ships could have been placed on the deck of one of them, with room to spare. Many, if not all, of the ships were commanded by Muslim eunuchs, who were considered to be the best navigators and sea commanders in the world. Between 1405 and 1433, there were seven naval expeditions that traveled out of Nanjing south to ports in what are now Vietnam, Indonesia, India, the Persian Gulf, the Red Sea, and the East African coast. The purposes of the expeditions were to extend the influence of the Celestial Kingdom, establish trading posts, and acquire some exotic items not available at home. Such items included animals like giraffes, zebras, and ostriches; jewels and minerals; and plants and seeds not known in China. One author[3] (Menzies, *1421*) goes so far as to claim that the fleets went to Australia and New Zealand, and even North America. Wherever they actually went, they were certainly capable of going anywhere, and it is likely that Zheng-He and other seamen were anxious to conduct extensive exploration.

But it did not happen. Instead, the emperor who had supported the expeditions died in 1424, and his successors were more occupied by internal concerns. The battles that ensued in the emperor's court pitted the explorers against the mandarins who controlled agriculture in the empire, and eventually the mandarins won out. The explorations came to an end, any dreams of world conquest were dashed, and this incredible fleet of ships gradually rotted in the harbor. China turned inward; it seemed that the outside world had nothing they would want. From the Chinese ruler's point of view, it's hard to fault this decision because the expeditions drained the imperial treasury without providing anything immediately valuable in return. Had they gone to South America via the Pacific Ocean, they might have discovered vast treasures of gold and other precious metals, and the world today might be quite different. But the ambitious explorers had no other source of support than the Chinese imperial court, so when that support disappeared, there was no plan B.

By contrast, Columbus lived in Europe, which was fragmented into many kingdoms that merchants, businessmen, and adventurers could travel between. This fragmentation was considered by Diamond and Landes to be a crucial condition for economic progress. According to Diamond, "Christopher Columbus, an Italian by birth, switched his allegiance to the duke of Anjou in France, then to the king of Portugal. When the latter refused his request for ships in which to explore westward, Columbus turned to the duke of Medina-Sedonia, who also refused, then to the count of Medina-Celi, who did likewise, and finally to the king and queen of Spain, who denied Columbus's first request but eventually granted his renewed appeal."[1] Columbus went on to discover the New World, and the news of his discoveries unleashed an orgy of enthusiasm for exploration.

This story shows how a multiplicity of opportunities in terms of contacts and potential financial support made a world of difference. There were many potential sources of support, and Columbus had access to them. It also shows how nondeterministic this process is. Suppose the king and queen of Spain had said no the second time: would Columbus have given up, and would that have changed history significantly? We'll never know, but if Spain's royal treasury had been more pressed for funds at that point in time, his second request could well have been turned down. In the case of the Chinese explorers, there was not even a second source of support, much less five of them.

Multiplicity Breeds Competition

At best, monopolies provide minimal encouragement for innovation. To illustrate this, consider a landowner who has a monopoly on flower sales in a town and sells the flowers though several flower shops. The managers of the shops will not be highly motivated to expand the selection of flowers since they don't get to keep the profits. This landowner might make some modest attempts to acquire new kinds of flowers, for example if some friends mention them to him, but that would be all. On the other hand, suppose there are two or more florists in town who profit from their sales and who buy their flowers from several nurseries or farms. If one florist hears about a new variety of flower, he or she will see an opportunity to sell some of the new flowers and make a little more money. The florists will strive to get the most customers and income by competing with each other on the basis of plant qualities—color, size, variety, shape, fragrance, and petal shape—and also on price, delivery options, and possibly credit arrangements. The result will be that customers will want to purchase new exotic plants, nurseries will be motivated to develop hybrids and obtain new kinds of flowers, and the sales volume will increase. Similarly, if there are two or more universities in a region teaching similar subjects, each will strive to be the best in some way. Their scholars will be the most recognized, they will publish the most books or articles, or they will get the most monetary support; or the student body will be the biggest, the tuition will be the cheapest (or the most expensive), or the athletic programs will be the most successful. By contrast, if there is only one university that is realistically available to potential students in the area, the motivation to improve the quality of teaching and research or to get the best scholars will be greatly reduced. *Competition provides an increased motivation for producing high-quality goods and services and striving to improve and expand them.* This is the basis of capitalism and its success as an economic system.

In modern Western societies, competition is given such frequent lip service that people take it for granted. However, even today many who advocate competition in the marketplace also actively oppose or resist when it becomes threatening to them. We have all tried to limit or suppress competition in our lives in one way or another—it's human nature. Companies and unions try to get their governments to impose tariffs on goods produced outside their countries. Companies seek to dominate their markets and often cross the line from competing fairly by providing better products to unfairly driving their competitors out of business. A classic example of this was the practice of John D.

Rockefeller of Standard Oil to offer lower prices for gasoline than his competitors in a region in order to drive them out of business. He would then make use of his resulting monopoly position to recover any losses and increase his profits. Another classic example, one that crosses societal boundaries, is the practice of dumping, where products (usually commodities) are deliberately underpriced, even sold at a loss, in order to drive competing companies out of business. This can be quite effective when the affected businesses are bankrupt and unable to reestablish their competitive positions once the "dumper" raises the prices.

Since competition is not in the immediate interest of any person or company, the only way competition can take hold in a society is for the forces of monopoly to be weakened. This weakening can happen when multiple power centers develop and when institutions are established to support competition. Such institutions have typically followed the emergence of distributed power within a society.

Competition and a multiplicity of power centers don't just benefit the economy of a society; they also create a favorable environment for innovation. In such a society, innovation can in turn increase economic activity. This mutually advantageous interaction between innovation and the economy is self-perpetuating because the resulting benefits become increasingly recognized by more and more members of the society, including the richest classes. This process stimulates innovation, providing new wants to be met and new opportunities for products and services. This is what happened in parts of the Western world. The story of how competition and a multiplicity of opportunities emerged is an interesting one.

Competition in History

The early autocratic civilizations were relatively small at first and were surrounded by wandering tribes, with whom they traded and fought. Eventually most of these tribes were conquered and assimilated into the larger civilizations. Within the autocracies commerce was dominated by centrally controlled monopolies, but between them a certain amount of trade occurred. For example, the Phoenicians established city-states around the south and east of the Mediterranean and maintained a brisk trade between them for several centuries. Once trade was multilateral and not just bilateral, a certain amount of competition occurred. When one society had goods to trade with different partners, it would trade the most with the customer who provided the most desirable goods in return. The laws of supply and demand began to take hold: if some goods were particularly desirable, the holders of the monopolies of those goods were motivated to increase their production. Conversely, when one society produced goods that it could no longer trade, and its merchants returned with them unsold, it reduced the motivation to produce them in the future. By the time of the Roman Empire, there were several cities where goods from different countries were traded and where money replaced barter. For example, in Jerusalem several types of coins were in use, notably Roman, Greek, Tyrean, and Jewish. The account of Jesus throwing out the moneychangers in the temple is believed to have involved these currencies for the purchase of livestock and other goods. This means that by that time markets existed where people from different regions haggled, bought, and sold goods.

Even before this, the Greek society had stadiums where citizens could watch athletes compete and amphitheaters where aspiring playwrights competed for prizes and for the approval of the audience. Craftsmen competed on the basis of the quality of their work. According to Cahill,[3] competition pervaded the lives of the Greeks. The freedom and empowerment of the citizenry made them self-consciously proud of their society. They and the other societies around the Mediterranean traded between each other, and a number of people not only were in contact with the other societies but also lived abroad for a while. For example, Pythagoras visited Egypt, India, and Babylon as well as cities around the Mediterranean to acquire their knowledge and wisdom, and he probably got most of his knowledge of geometry from Egyptian priests. However it came about, the Pythagorean theorem is generally considered to be an important mathematical innovation. People have always been curious about each other, and many who were financially able made contact with other societies. In other words, they took advantage of their access to multiple sources of knowledge and information.

The Greeks placed a high value on education and physical fitness. The sons of citizens who could afford it were taught reading and writing skills by tutors. Later they went to the gymnasium where they learned physical skills and received further training in reading and writing. They could then go to secondary schools where they could learn the science and mathematics of the time and debating and speaking skills. Advanced schools were established to educate scholars and other interested people. While poor children were not forbidden to get an education, they usually didn't have the means and were dependent on family and friends for any education they got. Slaves, girls, and manual laborers were not educated. Competition occurred between the tutors and the teachers in selling their capabilities; public education and public libraries, as we know them, would not be established until much later. The great libraries that did exist, for example at Alexandria, were primarily available to scholars by invitation. However, the notion that a broad-based education was a benefit to the society was now firmly planted.

The Roman Republic emulated much of the Greek experience and even set up libraries with separate Greek and Roman sections. However, it didn't have the same support from the entire population, and in many ways the society lapsed back into autocratic ways. Autocracies in India, the Middle East, and China continued to hew to their old ways, meaning that royal monopolies continued to dominate the larger economies. The multiplicity of opportunities in business and education in Europe diminished, and by the time of the fall of the Roman Empire, much of the progress had been lost.

By the 1600s considerable commercial traffic flowed between the various kingdoms, and merchants and traders were competing with each other. The merchant guilds formed their own courts to settle disputes, avoiding the oversight (and the associated delays and levies) of the local lords and monarchs. Towns were formed with citizens who were no longer bound to the feudal lords and estates. Guilds were established to provide artisans and craftsmen with some protection from the governments. The citizens were beginning to have a little more power over their lives, but not much. Businesses were created that involved piecework, wherein the efforts of individuals in the towns, villages, and countryside were directed and collected. Later these were replaced by centralized factories, which more closely controlled and standardized fabrication practices. Once businessmen started competing to provide clothing, food, and other goods, the stage was set for technological innovation.

The changes were most evident in the Mediterranean societies, in particular the city-states of what today is Italy. The Venetian Republic was at the center of trade and for a few hundred years was the most populous and powerful city in Europe. While its government was a plutocracy, political power was quite widely distributed relative to other societies. There was some social mobility among the families, and the lower classes were relatively well-off by the standards of the times. It comes as no surprise, then, that many improvements in navigation and shipbuilding took place there. Other nearby city-states were the hubs of a large trade network: Genoa, Florence, and Milan were financial centers that were the homes to large organizations acting as lending banks, investment banks, and import/export companies. In these city-states, almost half of the boys attended schools, and many went on to abacus schools, where they learned arithmetic using Arabic numerals, including zero, and accounting methods.[4] They formed a new profession that kept the records of the large companies, and they were sent to branches in Rome, London, Dublin, Flanders, Paris, Bordeaux, and other cities around the Mediterranean and in the Middle East. Business was conducted in Italian. The city-states were the poster children of the times, a testament to the benefits of a weakened class structure and widened trade. The rich families got to wear clothes with exotic new fabrics, experience new foods and spices, and, with the invention of the printing press, avail themselves of the knowledge and wisdom of the past. Many people on the lower end of the economic scale were considerably better off than their counterparts to the north.

The rich families of the city-states considered themselves to be enlightened, the torch-carriers of Greek wisdom. They competed with each other to demonstrate this by sponsoring scholars, scientists and other thinkers—Galileo was sponsored by his student Cosimo Medici, and Michelangelo was invited by Lorenzo to be educated with the Medici children.[5] It became a matter of pride to have a university in the city, so the university at Bologna was soon followed by dozens of others. As the rich European families to the north heard about these, they also wanted to lay claim to being enlightened, so universities were soon established in France, Britain, and Spain. The Church also supported the universities, seeing them as a means of better educating their young clerics. While they were primarily for educating clergy and the sons of the ruling classes, their administrators were given some latitude, and they developed some independence in their teachings. They, too, began to compete with each other, and this competition grew over time.

The practice of issuing exclusive rights for certain commercial activities was used by the monarchs of Europe to collect taxes and reward their favorites at court. Monopolies were granted even on such items as playing cards and mail delivery. These rights, called patents, did not initially have expiration dates, so after a generation or two they resulted in a skewed set of incentives; changes brought on by trade made many patent terms either too costly or meaningless. After they were terminated by royal decree, usually several competitors were able to supply the same goods at much lower prices. When limited-time patents were introduced and other antimonopoly laws were enacted, competition became more common.

As the merchants became more powerful, they found ways to get their local governments to support their interests. They formed guilds that established monopolies in the towns and countries. Craftsmen formed guilds that established monopolies on their products. The game was to keep new ideas secret, however, rather than to make them widely available. As populations grew and commerce increased, Europeans looked outward for sources of wealth, and the age of exploration began. This was followed by a period of colonization, whereby they took over much of the productive capacity of the world, with some old-fashioned looting and pillaging thrown in, plus the enslavement of people on a gigantic scale.

With colonization came mercantilism. The idea of mercantilism is that a country's wealth is built up by exporting more goods than are imported, and making up the difference in gold and precious metals. This fit well with colonialism because the colonies provided much more of the goods to be exported than the mother countries could produce themselves. The health of the economies of the colonies was of little concern. The monarchist governments and the merchants and landowners cooperated closely—governments used their armies and navies to protect the merchant groups, and the tariffs and taxes filled the government coffers. Exports were often subsidized, and tariffs and even prohibitions were placed on incoming goods. Prohibitions were also placed on the export of gold and silver. To the extent possible, raw goods were to be imported, and finished goods made from them were to be exported. The goal was to maximize production and minimize consumption, especially of imported goods. The cost of workers was to be kept low, and there was no incentive for their lives to be much above subsistence, as consumption by the multitudes in the lower classes was seen as something to be minimized. The people who advocated this doctrine saw trade as zero-sum game, in which one party gained at the other's expense. Simple arithmetic shows that if the mercantilist countries traded only with each other, they couldn't all achieve the goal of keeping exports above imports—there had to be winners and losers.

Colonialism and mercantilism did not lead immediately to the appreciation of competition—quite the opposite. A story that demonstrates this is that of the origin of the Dutch East India Company.[6] In 1594 the Company of Far Lands in Amsterdam sent two fleets to some spice islands within an archipelago of Indonesia to obtain some pepper to sell in European markets. The first fleet brought back enough to break even, and the second generated a fourfold profit. This led to the founding of several other competing companies anxious to cash in. Unfortunately for them, the result was that the increased demand for pepper raised the price at the spice islands, and the increased supply resulted in a reduction in the prices charged in the European market (surprise!), so no one did very well financially. This intercompany rivalry was considered quite problematic, so the Dutch East India Company was established in 1602 to eliminate this destructive competition, and it was given a monopoly position for twenty-one years. So much for free enterprise in those times.

The mercantilist doctrine did increase production worldwide, and it had effects not envisioned by its proponents. The mercantilists envisioned a kind of plutocracy wherein the main features of the old autocracies were maintained but where the autocrat was replaced by an enlightened group of noblemen. To work as they envisioned, it required a large population of passive subsistence-level workers in the fields, the mines, and the factories and to be servants to the rich. Unfortunately for them, but fortunately for the world, the situation had changed. The workers had options and influence that their predecessors didn't have.

As parliaments grew in power in Europe, the monopolies granted by monarchs were eliminated, and merchants and other businessmen increasingly competed with each other. To gain a monopoly position, a person or company had to convince parliaments and government bureaucrats that it was in the national interest. This was much more difficult (but still possible, as experience showed) than convincing a king that it was in his best interest.

With competition taking place both within and between countries, the modern free-market concept emerged, with its emphasis on laws that maintain a level playing field. It became widely recognized that the society is best served when a competitive environment is maintained: goods are cheaper, people become more prosperous, and—what is most important for this book—competition stimulates innovation. A competitive market for a product provides motivation to lower production and shipping costs. Innovations that can achieve this include better or cheaper materials, better manufacturing techniques or assembly methods to reduce labor time and improve quality, better machines to reduce labor costs and improve quality, new packaging or shipping concepts to reduce delivery costs, and improvements in the next generation of the product to make it lighter, more reliable, more attractive, easier to use, smaller, easier to install, or cheaper. All these improvements require innovative activity. Some of them result in the establishment of new industries that make possible similar improvements for other products.

In the last two hundred years, we have seen an accelerating growth in the multiplicity of opportunities in many areas of human life that are affected by innovations. One has been in higher education: as transportation improved, it became easier for students to go farther from home to attend colleges and universities, especially in the United States. This brought subcultures together, resulting in competition between institutions of higher learning and exposure to different ideas, always stimulants to innovation. Later the same factors made it possible for more students to study abroad, bringing societal cultures together, with even more stimulation of the innovative spirit.

Another area of life that has experienced increasing multiplicity is employment. As economies have grown and more companies have come into existence, the workforce has become more mobile. Improvements in transportation make more work choices possible farther from home.

Advances in transportation have made it possible for consumers to travel farther from their homes to stores with a wider selection of products, giving them a multiplicity of product choices. The internet has made possible online purchases of both new and used products, which has expanded these options considerably. Businessmen have taken advantage of these capabilities to greatly expand both their supplier chains and their sales bases, both domestically and abroad. Plus, with the huge increase in the availability of information, they can more readily compare capabilities and evaluate new business opportunities. Multiplicity breeds multiplicity.

15. Three Great Innovators

Galileo Galilei, the genius with blue stains in his white collar;
Philippe de Girard, a forgotten mega-inventor;
Thomas Edison, of course; others

We have learned that for innovation to flourish, the societal conditions relating to fear, information, rewards, and opportunities must be present. When they are, curious and innovative people will find ways to pursue their passion for critical thinking, challenging existing concepts, and proposing new ideas and inventions. While we all have some innovative spirit and talent, some individuals are, for whatever complex of reasons, more highly motivated than the rest of us.

There have been many great innovators throughout history, and each lived a unique and fascinating life. One of the great pleasures in life is reading about them, their motivations, the obstacles they faced, the support they got from their colleagues and friends, and their successes and disappointments. They were very diverse. Some came from rich families, some from poor ones. While some were highly educated, many were less so.

I have selected three individuals, whose lives collectively spanned four centuries. They share several traits typical of innovators: extreme curiosity, high intelligence, incredible perseverance, and a willingness to work hard. Above all, they had a passion for developing new ideas or products, and that's how they wanted to spend their lives. They were not driven primarily by a desire to become rich, just to become prosperous enough to continue what they wanted most to do: ponder, create, invent, innovate. This is evidenced by the fact that they continued to innovate throughout their lives regardless of their financial conditions.

I have chosen Galileo Galilei of Tuscany, Philippe de Girard of France, and Thomas Edison of the United States as representative of innovative personalities. Galileo was the most white-collar of the three, but he was not a snob about it, and he spent much of his career performing experiments. He actually ran a small business while he lived in Padua designing and building telescopes and had craftsmen do the fabrication, but his driving interest at the time was using his telescopes to point at objects in the sky, and he used his observations to challenge conventional wisdom. Philippe de Girard was one of the most inventive individuals ever, one that few of us have ever heard of, and he was the classic case of an inventor who had very little business sense, lost a fortune, and died a pauper. Thomas Edison had a hands-on knowledge of markets and ran several businesses, but his love was inventing—and invent he did, right up to the end.

As you read about these great men, you will see how the presence or absence of the five essential conditions for innovation played out in their lives.

Galileo Galilei (1564–1642)[1]

Galileo is best known to us as the scientist who looked into the skies with a telescope and, based on what he saw and inferred, showed that the solar system, rather than having planets with complicated motions around a stationary earth, was in fact a relatively simple system of planetary spheres traveling around the central sun in circular orbits. While we now know that the planets aren't perfectly spherical and their orbits not quite circular, these new discoveries had profound philosophical and religious implications. He was fortunate enough to be recognized and respected in his own time, and he was able to spend most of his life productively. He is recognized as the person more than anyone else who put scientific experimentation on a sound footing.

Galileo grew up in Tuscany, the son of a noble family that had lost much of its fortune. His father was a musician who developed a scheme for tuning lutes that used a tempered scale and who eventually wrote a book on the subject. After attending grammar school, Galileo was trained by a monk who taught him Greek, Latin, and logic. Later he attended the University of Pisa, where he studied medicine and mathematics. He was exposed to Aristotle's theories of perfect forms in nature, the mathematics of Archimedes, and the geometries of Euclid. Impressed by the intellect of these ancients, he decided to concentrate on mathematics. He conducted independent research at home and wrote up his results. He was able to use these manuscripts, which demonstrated his talent, to obtain an appointment as lecturer at the University of Pisa.

Galileo did not stay at the University of Pisa very long. He got in some trouble by not wearing the required black togas, and by writing bawdy poems that made fun of his colleagues. More important, he found that the other faculty members were unwilling to critically examine the assertions of the ancients, something he enjoyed doing. He made repeated attempts to obtain a position at the University of Padua in the Venetian Republic, and after three years he was successful in getting a faculty position there at the age of twenty-eight. The Venetian Republic gave the University of Padua wide intellectual freedom, and it hired the best teachers in Europe.

Venice enjoyed considerable political independence from Rome, which meant that the Vatican had less influence on philosophical thought than in other cities of Italy.

He was wildly successful at the University of Padua and conducted much of his best research there. In 1597 he invented a computing device that was a kind of slide rule that he called a "compass." It could determine square and cube roots and did calculations useful to military strategists and shipbuilders. He built a few by himself and sold them, and eventually he hired a craftsman to keep up with the demand. His shop then not only became a source of income but also enabled him to fabricate highly accurate instruments and equipment.

In Galileo's personal life in Padua, two major events occurred: he accidentally inhaled a considerable amount of poisonous gas from a vent from a cave, and he fell in love. The poisonous gas resulted in frequent bouts of pain that plagued him for his whole life. The object of his affections was not a woman of nobility, so he didn't marry her, but she became his mistress. They had two daughters and a son. He eventually sent his daughters to a convent, to which he contributed financially. He was close to his oldest daughter, Virginia, and he maintained a correspondence with her the rest of his life, which is the subject of a delightful book, *Galileo's Daughter,* by Dava Sobel.

Another important event happened during this time: Giordano Bruno, a friar who promoted the viewpoint that Copernicus's theory that the earth rotated and orbited around the sun was not just an intellectual hypothesis but represented reality, was burned at the stake by the Inquisition for heresy in 1600. This must have made quite an impression on Galileo, who was thirty-five at the time, but apparently not enough of an impression for him to avoid his own later interrogation by the Inquisition.

Galileo had a wide variety of interests, and he was lucky enough to live during a period when not one, but two, supernovae occurred. The first occurred when he was small, but the second occurred in 1604 and was visible for several days. Since astrology was still believed by most people to be valid, one can only imagine the excitement that this supernova produced, and people everywhere speculated over its significance for human events. According to Fermi and Bernardini,[2] it would change the course of Galileo's life. He began to make systematic observations and realized that the sky had critical information to offer about the workings of the universe. He correctly concluded that the new object (we now know that it was an exploding star) was as far away as the stars, which contradicted the accepted Aristotelian theory. That theory held that the "celestial sphere," in which the stars were located, was unchangeable; thus, it couldn't contain a new and changing star, leading to the erroneous conclusion that the event must have occurred close to the earth. This gave Galileo more motivation to challenge the ancients.

In 1609 he heard about a new device that was purported to make distant objects appear closer. He figured out that if the device did what was claimed, it would have to have more than one lens, one convex and the other concave. Following this reasoning, he constructed his own telescope. Subsequently, he experimented with different lengths and diameters, developed a theory of operation that involved the determination of the curvature of the lenses, and developed methods for grinding and polishing the glass lenses to a degree that was not matched by others for decades. He was able to achieve a magnification ratio of 33:1 before distortion set in. He immediately recognized the military advantage it would provide. By pointing it out to sea, the observer would be able to spot enemy ships hours before the enemy could see the observer. He prepared a dramatic demonstration at the top of the Campanile, the bell tower in San Marco Square in Venice, to which a large number of senators, noblemen, and academicians were invited. They were all able to look out over the harbor through the telescope, and, as you would expect, they were blown away by what they saw. Later he gave it as a present to the highest-ranking prince in Venice, who decided that Galileo should be given a permanent position with a generous annual salary.

As Fermi and Bernardini correctly noted,[3] Galileo was a true scientist in that, while he basked in the glow of his acclaim, his primary interest was to use this incredible new tool to address the great questions of science. He pointed his telescope at the sky, and in so doing he became the first person in the world to see features of the universe not visible to the naked eye. It must have been one of the most exciting experiences of any man, before or since. He felt that God had chosen him to be the recipient of secrets of creation that no human had ever seen before, and he often expressed his gratitude to the Deity for this in his writings.

When he looked at the moon, he recognized that he was seeing mountains, plains, and valleys, just like those on earth. When he looked at Jupiter, he found small pinpoints of light around it that moved noticeably even during one observation session. He correctly concluded that they were moons of Jupiter, ones that orbited that planet just like the moon travels around the earth. This had profound implications for his arguments against the Aristotelian model of the universe. He later realized that the motions of the moons were repeatable but different from each other, so that they could be used to synchronize the time of events in different parts of the world. That is, the pattern of the points of light, which could be seen all over the world at the same time, could be used to establish the day and hour, and by noting the instant that a particular moon passed behind Jupiter's disk, the time could be recorded quite accurately. That instant would be the same wherever it was observed from earth. This technique was actually used later as a practical means for determining the differences in longitude at different places on earth. It should also be noted that Galileo, who was interested in the measurement of longitude, recognized that the ability to carry time from one place to another would theoretically enable sailors to determine their longitude at sea.

Galileo also discovered that Venus had phases like those of the moon and found that Venus changed in appearance at different times of the year as it came closer to the earth. No doubt covering the telescope lens with heavy paper, he was able to discern sunspots on the sun (which others had seen earlier without the use of telescopes) and tracked their motion, which led him to assert that the sun rotated around its axis like the earth. Since the telescope concentrated the light from the stars, he was able to see the myriad of stars and heavenly objects that were not visible to the naked eye. He published these observations in letters and in March 1610 finished his book, *The Starry Messenger*, which described his observations. His publications made him famous and established his fame worldwide—his book was even translated into Chinese within five years.

Unfortunately for Galileo, he was not as politically astute as he should have been. With the freedom he had at Padua came burdensome obligations. He was required to meet with students, faculty, and visiting noblemen, who understandably flocked to his laboratory. With each new publication, he found himself obliged to spend much of his time explaining his work to people, time that he would rather have spent on his research. He finally appealed to Cosimo Medici, his most influential student, to provide him with a position back in Tuscany. After eighteen years at Padua, he moved back to Florence as personal philosopher and mathematician to Grand Duke Cosimo, with no requirements to teach students or meet the public. Giving up the messy freedom of Venice for the seemingly quiet haven of Florence was to have momentous consequences.

Initially everything went beautifully. He went to Rome in 1611 with his telescopes and observations, where he was treated as a hero by cardinals and Vatican officials. He was elected to the Academy of the Lynxes. Believing that the intellectual climate and his now-famous observations and discoveries made it acceptable to challenge Aristotelian notions, he set out to make the Copernican system acceptable to the theologians. However, some of his enemies made accusations to the Inquisition; this required the Church to respond, which it did in a decree in 1616. The result was that Galileo's friend Cardinal Bellarmine told him publicly not to publish anything that claimed that the Copernican theory was anything beyond an interesting mathematical hypothesis. Years later, in 1623, another of Galileo's admirers and friends became Pope Urban VIII, which emboldened him to try again to advocate the Copernican theory as reality. Believing that he had the new pope's blessing in the endeavor, he wrote the *Dialogue*. In this, he had agreed to state the Copernican theory as scientific hypothesis, not as reality. However, he chose the character Semplicio to make this statement near the end of the book. The book had made Semplicio look silly, hence his name, and it's possible the pope took this as an insult. In any case, the pope turned against Galileo, he was hauled before the Inquisition, and the famous scientist was forced to recant his assertions in 1633 or face death.

Being prohibited from publishing and having his books removed from print, Galileo still was given protection by sympathetic clergy and continued his experiments and writing. He completed his magnum opus, *Discourses and Mathematical Demonstrations Concerning Two New Sciences Pertaining to Mechanics and Local Motions*, commonly abbreviated as *Two New Sciences,* which contained his research over a forty-year period. A few years before he died in 1642, and probably sensing that he had little time left, he quietly sent the book to a friend in Venice, who sent it on to Louis Elzevir, a publisher in Holland. Beyond the reach of the Church in Rome, the book was published and sold throughout Europe. The *Dialogue* was similarly published during his lifetime. The general prohibition against discussing the Copernican system was finally lifted in 1757, but it wasn't until 1835 that his books were finally removed from the Church's index of prohibited books.

Galileo was supremely confident in his own capabilities and opinions. He was quite caustic, as evidenced in the *Dialogue*, and enjoyed attacking other people's ideas. When he taught in Venice, he would often describe theories of the time with such compelling arguments that the class would be convinced they were correct; then he would proceed to methodically tear the arguments down and show that they were wrong. In the *Dialogue* and *Two New Sciences*, he used the device of a conversation between three people, with one protagonist representing Galileo's views and presenting the more powerful arguments and another representing the old-world views. The third person played the role of foil and arbiter, asking questions that an astute student might pose. He wrote these in Italian rather than Latin because he wanted to reach a larger audience than the scholars who had learned the ancient language.

One of the problems in his research was that the mechanical clocks of the time did not provide enough resolution and accuracy for his work. So in true scientific spirit, he took the occasion to invent two better methods. One used water, the other pendulums. The water method opened and closed a calibrated flow of water through a pipe—the volume of water was an indication of the time elapsed. The other method used manually pushed pendulums that, once in motion, gave excellent time marks over short periods. Late in his life he invented a method by which a weight could be used to periodically give a little push to keep the pendulum going. Blind by that time, he described this design to his son Vincenzio, who drew up a picture but didn't actually fabricate the device. Christian Huygens, the famous Dutch astronomer, built the first pendulum clock in 1657, giving credit to Galileo's work.

In his studies of mechanics, Galileo had to address the problem that dropped objects traveled so fast that making accurate measurements was difficult. He solved this problem by setting up an inclined plane and letting cylindrical and spherical objects roll down it. This slowed things enough that he could measure time intervals, compute proportions, and derive principles. One important example was that a falling body falls exactly four times as far in two units of time as in one unit of time and that its speed increases by a fixed amount for each unit of time. He understood the concept of instantaneous velocity and other aspects of calculus, although he didn't develop its mathematical structure. Probably because of the difficulty of accurately measuring small durations of time, he confined his research to principles that could be described in relative time terms. He accurately claimed, with experimental proof, that motion could be separated into three directional components. He showed that in the absence of friction (demonstrated with objects with very little friction, like smooth, hard balls) a freely moving object will continue to move horizontally at a constant speed, while the vertical motion will respond to the pull of the earth by increasing its speed linearly with time. Combining the components, he said the motion of a projectile would be that of a parabola; he also knew and stated that the parabolic description was accurate only when the trajectory length was small compared with the radius of the earth. He was even astute enough to recognize the effects of friction. All this occurred in the early 1600s.

As with many innovators, Galileo's supreme self-confidence occasionally made him overreach, that is, to not fully think through or experiment through some of his assertions. His explanation of comets was wrong (he claimed they were optical phenomena and not heavenly bodies), even though his book *The Assayer (Il Saggiatore)* was considered one of the best polemics of its time. His theory of tides was also wrong, as he denied the role of the moon's gravitational pull and attributed the tides instead to a combined effect of the earth's rotation and movement around the sun. This does not detract at all from his incredible accomplishments; rather it is a reminder that even the greatest minds are not invincible. Instead of assuming that an assertion is true just because some great thinker claimed it, a person should be able to reexamine and criticize it without worrying about being accused of disrespect, or even heresy.

It is important to recognize that Galileo was not motivated by profit, but rather by a desire to understand the physical world. He could have made a reasonable living manufacturing his compasses, pendulum clocks, and telescopes. Instead, most of his inventions were attempts to better measure physical quantities of nature and observe the physical behavior of matter. Also, as with all inventors, he was driven not by an urge to understand only for his own edification, but rather to share this understanding with the world at large. That is why he took the considerable risk to sneak out *Two New Sciences* to get it published.

The five essential factors I have identified for an innovative society and how they operate are illustrated quite graphically by Galileo's experience. Living when and where he did, he had access to the knowledge of the times. Being the son of a nobleman, he obtained educational guidance as a youth and learned the Greek and Latin essential for accessing the information he needed for his eventual research. His training in medicine and mathematics, which really included the physics of the day, gave him the tools to challenge the geniuses of antiquity. Multiplicity of opportunities was essential in many ways. First, he had the choice of several universities where he might find a position: there were about seventeen universities in what we now call Italy, and at Padua he found an environment friendly toward challenging even the authoritative Greeks. Second, even with his entanglement with the Inquisition, he had friends in the clergy who did their best to encourage and protect him. Third, he had a network of contacts with astronomers, scientists, and philosophers. He was able to learn about new inventions, such as the rudimentary telescope in Flanders, that were crucial to his success and development. He received a number of societal rewards: a lectureship at Pisa, a lectureship followed by a lifetime appointment at Padua, and a position with Grand Duke Cosimo di Medici. The monetary rewards for his inventions and his election to the Academy of the Lynxes also encouraged and motivated him.

What about freedom from fear, however? After all, when Galileo was thirty-five years old, Fra Bruno Giordano was burned at the stake for advocating similar ideas. Wasn't it true that he caved in to the Inquisition and stopped advocating the Copernican planetary model? Actually, by the time Bruno was killed, Galileo was accustomed to being able to think and publish as he wished. Fear of the Inquisition made him cautious about how he proceeded, but he continued to think, write, and perform his experiments. Contrary to what many people think, he was not totally cowed by his experience with the Inquisition, as his publication of *Two New Sciences* demonstrates. The Church was unable to keep Galileo's ideas from spreading throughout the world, although it did a lot of harm in Catholic countries by preventing the people from reading his ideas and those of other thinkers.

Galileo is generally credited as the first to put science on a firm experimental basis, where one always looks for methods of testing theoretical claims and where those that pass such tests are given greater credibility than those that haven't been tested. Observations trump the claims of experts. He was one of the greatest geniuses of all time.

Philippe de Girard (1775–1845) [4,5]

Philippe de Girard is one of the most interesting and amazing people in the history of innovation, but few people have ever heard of him, even in France, his beloved country. I discovered him quite by accident on a trip to Provence. My wife and I had met a couple whose family had a summer place in Lourmarin, and they offered to rent it to us for a week. While reading through the *Michelin Guide*, I came across a single paragraph description about Philippe de Girard that made me realize he was the perfect archetype of a truly single-minded innovator—his capacity for inventions was limitless, while his capacity for making money from his inventions was almost nonexistent.[6] The account characterized him as having little practical common sense, but this is not a fair assessment, as I will show. He was very practical, and his inventions worked very well. He just had no ability to market and sell his inventions for profit. This is a very common trait of innovative people. In fact, there are relatively few people who have both a talent for invention and a talent for making money from their inventions. Sometimes this is because of their lack of interest in the details of setting up a business and making it succeed, and sometimes this is because of a total lack of business talent; the societal environment also plays a crucial role. Thomas Edison comes to mind as a relatively rare exception, and he will be discussed later in this chapter.

Philippe was born into nobility in Lourmarin in 1775. His inventions spanned a wide range of technologies; they included a hydrostatic lamp, an achromatic telescope, a steam-driven machine gun, a hydrochloric acid factory, a machine for solving equations, a piano that could make vibrato sounds, and a chronothermometer, a device that measures temperature by measuring variations of time with changes in temperature. But his most important invention was a linen-spinning machine, and the story about that is incredible. During his lifetime he was in debtor's prison at least twice, and was bailed out by his brothers. Many people around him made fortunes off his ideas, but he died a pauper in 1845.

Growing up at the time of the French Revolution was difficult because his family backed the royalists and took part in the fights. At the outset of the revolution in 1789, he went to study medicine in Montpellier, in the south of France. While looking over the sea of Pezenas, he came on the idea of using ocean waves as an energy source—this at the age of fourteen! [6] As the political situation worsened, the family fled by boat to Minorca in 1793, then to Livorno, Italy. While there, Philippe developed a steam-driven engine for a local soap factory, a process for reproducing small statues, and a process for engraving hard stones. In 1794 the family was allowed to return to France. Philippe went to Marseille, where by 1803 he had obtained patents on three more inventions, two relating to ships and one for a musical organ that could crescendo and decrescendo without changing the quality of the sound.

In 1805 he moved to Paris, where at that time there was a lot of interest in replacing animal energy with machines. A four-year exposition was held that offered prizes for useful inventions. Philippe achieved quite a measure of fame among his peers by presenting four inventions there: (1) a sophisticated steam-engine design, (2) a demonstrated technique for applying varnish to sheet metal plates, (3) an achromatic lens, wherein liquid was used inside the glass to suppress the color-separating prism effect of conventional lenses, and (4) a hydrostatic lamp. He was awarded a gold medal in 1809 and given lavish praise for his efforts, but he received no money and no support for producing any of these inventions on a commercial scale.

The hydrostatic lamp was his pride and joy. It had two major advantages over the conventional lamps at the time: (1) the hydrostatic feature ensured that the burn rate and thus the luminance of the lamp were more uniform (i.e., the flickering was greatly reduced), and (2) the efficiency and the brightness of the lamp were greater. In 1807 the Grand Duchess of Berg had her apartments brightly lit with a number of these lamps, and Empress Josephine saw them and expressed her amazement and admiration. This got back to the de Girard brothers, who immediately fabricated a pair of lamps and delivered them to the empress.

However, for some reason, Philippe and his brothers didn't follow up this wonderful design by setting up a business. Instead, a low-level commission agent, who visited the de Girard household and observed the lamps, returned to his attic and, with help from some of his children, made a practical copy that he attempted to build and sell in France. When the French authorities found out, they tried to shut him down. However, he moved to Belgium with his family, set up a business there, and became quite rich.

It is the story of the linen-spinning machine, though, that demonstrates the role that sheer luck, good or bad, can play in life's major events. If certain events had happened a year later, Philippe de Girard and his family might have been financially secure, and he would have been free to engage his inventive mind for the rest of his life. His name today might be as well known as Thomas Edison or Benjamin Franklin, or he might have squandered his fortune on another business venture that went sour—we'll never know.

In any event, the story is this.[7] On May 12, 1810, Napoleon issued a decree that promised a million francs to the inventor who could demonstrate a practical technique for spinning linen, which comes from flax fibers. It was at lunch on that day that Philippe's father received the *Monitor*, a newspaper that contained the decree. When he saw it, his eyes widened, and he said to his thirty-five-year-old son, "Philippe, look at this!" When Philippe saw it, he realized that this was a golden opportunity to solidify his family's fortune and establish himself as a major inventor in France. He gathered up some flax fibers and thread and a magnifying glass, locked himself in his room, and proceeded to experiment. He washed the fibers in water and other solutions and tried everything he could think of to understand the properties of the material and to figure out how to spin it. Flax fibers are extremely fine and difficult to work with, which is why there wasn't already such a machine. He spent the rest of the day and all that night working away, finally emerging the next day at lunch. Triumphantly, he announced, "The million is ours; I just have to do with a machine what I can do with my hands, and we will have a working machine!" He feverishly continued to work out a design, and two months later, in July 1810, he received a patent, which established him as the prime candidate for the prize. The patent description was very thorough. It consisted of an analysis of the organic structure of the flax, a description of the process employed to separate the filaments, and a description of the apparatus needed to form the threads—all of which happened in less than one month after the publication of the imperial decree.

However, he was not immediately awarded the million francs on the basis of the patent; instead, he was asked to demonstrate a prototype before he could receive the reward. This required him to spend a considerable amount of his own money. He hired a clockmaker with a reputation for precision work to help him develop a small prototype. He then took out two more patents for improvements in the design, one in January and one in May of 1812. Later he set up two workshops in Paris to produce the linen on a commercial scale. The thread produced was said to be so light that a seventeen-mile-long thread would weigh a pound. Napoleon himself visited his workshop and in May 1812 wrote to his minister of commerce and told him to put together a jury to examine the linen-spinning machine and to award the prize if Philippe's technique was judged to be feasible. It looked like Philippe's fortune was assured.

Then a series of disasters struck. His family's primary sources of income, businesses that produced salt, soda, and soap, were devastated by new parliamentary laws that allowed cheap foreign imports of these materials into the country. Philippe was thrown in debtor's prison for a time because he couldn't repay his loans. Napoleon's ambitions were thwarted by an allied coalition that included Austria, Hungary, Prussia, Russia, and Sweden. By early 1814 the allied forces were threatening Paris. With his last piece of gold, Philippe bought a rifle to help defend the city, a naive move and to no avail—Paris fell, and his spinning machines were dismantled so they wouldn't fall into enemy hands. Napoleon was exiled to Elba, and his government was replaced by the monarchy of Louis XVIII. In March 1815 Napoleon returned briefly to power, only to be deposed for good after Waterloo in June. Philippe appealed to the new government to honor Napoleon's promise, but without success. The final blow came later in 1818 when a government report declared that his spinning machines "did not merit the honor of being placed at the Conservatory."[8] By that time Philippe was deeply in debt, as was his family, who had invested in his enterprise.

Philippe was forced to stave off bankruptcy when the government changed hands. In October 1815 his friend Constant Prevost introduced him to some Austrian officers who told him that his inventions had attracted the attention of the Austrian emperor and that he was invited to equip Austria with his linen-spinning machines. He hesitated, saying in his naive way, "My invention belongs to my country." With creditors chasing him and no other source of income in sight, he was finally forced to accept the offer; he also knew that if he was in Austria he couldn't be put into a French debtor's prison. He took some of his machines with him and undertook to set up a factory at Hirtenberg. For some reason, it took two years to build a working machine there, but an imperial commission recognized the value of the machines, and factories were set up in several districts. One would think that he would have been well set up with this arrangement, but he was again able to snatch defeat from the jaws of victory—he took out a large loan, defaulted, and had to sell some of his machines to pay it off.

Probably, he needed the money to fund yet another invention, building on his earlier work with steam engines. His new ambition was to build the first steamboat on the Danube. He came up with an arrangement of tubes that countered the impact of the explosions common to such engines. He then entered a competition at Presbourg, and he had the satisfaction of demonstrating his genius to an appreciative crowd by executing a complete circle on the river. Again, no investors stepped up to build steamships based on his design.

For the next few years, Philippe went through bouts of illness that left him incapacitated. In 1824 he went to the spa at Baden, and there he made the acquaintance of the Russian Prince Lubecki, who was the royal minister of finance in Poland. Philippe, at ease with nobility and a gifted talker, charmed him with his imagination and enthusiasm, so much so that in 1825 the prince offered him the position of engineer of Polish mines, which he accepted on the condition that he got to keep his French citizenship. The contract guaranteed a generous annual salary, housing allowance, and travel expenses for a period of ten years. At last, perhaps, our hero could relax. But having heard that the family home in Lourmarin was about to be seized by creditors, he borrowed against his salary to pay it off. So again he was in debt, now in Poland as well as France.

He was sent to England in 1826 to recruit workers for the mines in Poland and to check out the progress in linen-spinning machines. There he found two large factories and twenty small ones that were actively producing linen. While he was there, he managed to read the British patent of 1815; he then wrote his friend Prevost that he had the pain of seeing his own drawings in it. While he was in debtor's prison, his French associates had made copies of his drawings and sent them to England, where they took out a patent with a person named Horace Hall. They were well rewarded for this effort. The British engineer who implemented the design and the French investors became millionaires. Philippe made a public complaint and took his case to the British government. Responding to his complaint, his majesty's government ruled that the idea behind the invention was indeed that of de Girard. This admission gave him some personal satisfaction, but as usual he was not able to get any financial compensation.

After returning to Poland, he developed a new design for a zinc furnace and developed an ingenious means of making a cheaper fuel. His sponsor, Prince Lubecki, stated publicly that he had saved more money in two years at a single factory than the worth of de Girard's salary for ten years. His most significant contribution, however, was to set up a whole industry of linen-spinning machines in an area called Ruda Guzowska, near Warsaw, an area suitable for the growing of flax. About five hundred families moved in to support the industry, and it became the center of linen spinning for several decades. The town was named Zyrardów (Girardov) in his honor. For several years, Philippe enjoyed the admiration and respect of the people, and he had a source of income. He was invited to the homes of the nobility, most notably the home of Grand Duke Michel. There he was introduced to Czar Nicholas I, who pinned on Philippe's chest the Cross of Stanislas.

In spite of his ease with nobility and royalty, Philippe was not politically astute, as we have seen. Sure enough, two jealous German nationals who came to the country after the revolution of 1830 convinced the new authorities to force Philippe to retire as engineer of mines.

He then devoted his time to inventing again. He perfected a technique for forming wooden rifle stocks, which he tried to get patented and produced in France, unsuccessfully. He took out a patent on improvements in linen treatment, involving an apparatus for cleaning and combing linen. In 1835 he was hired as a consultant to the Bank of Poland, where he was given a small but adequate salary. He developed a new technique for the treatment of beets and established a sugar beet factory in Gouzow. Upon visiting the home of an industrial director on an inspection tour, he amused himself by attempting to fix the piano of the daughter of the house. In the process of learning the intricacies of piano design, he figured out how to produce two sounds an octave apart with a single keystroke, and he subsequently invented what he called the “octave piano.” At some point he also invented a piano with tremolo, a feature that varies the frequency to give a “wavy” feeling to the tone. For the bank, he developed a chronothermometer, which continuously displayed a plot of the outside temperature during the previous twenty-four hours. He also developed a meteorograph, which recorded the temperature, humidity, atmospheric pressure, wind direction, and wind speed. He didn’t attach a lot of importance to these, probably because he considered them to be rather obvious extensions of current practice, but the czar gave him a diamond ring for these efforts.

Sometime after Philippe left France, two industrialists used his linen-spinning inventions to set up the region of Lille as the first center of the spinning industry in France. That city belatedly recognized his contributions and still has memorabilia in his name, including a street named after him.

These inventions were not enough to pay all his French debts, which dated back to 1813. Going back to France meant risking debtor’s prison. He longed to return to France, and this took a toll on his health. Upon learning of the state of her uncle, Philippe’s niece (a countess) and her daughter went to Poland and pleaded with Philippe to present some of his inventions at the Universal Exposition of 1844 in Paris. He eventually consented and attended the exposition, where he was received with great honors and acclaim by his compatriots. It was a bittersweet experience, however, because he also saw on display several inventions of his own that were attributed to others.

He hid outside Paris for a few months, finally returning to Paris on his seventieth birthday on February 1, 1845. At that point, he could no longer be imprisoned for his debts. He died in August of that year. His niece and legal heir, Contessa de Vernède de Corneillan, and her daughter were awarded a small pension.[9]

What are we to make of this man? He was indeed a tragic figure, and his story could be the basis of a Greek epic poem. He had immense energy, intelligence, ambition, persistence, and charm. He started with a fortune and connections, did his best to bring honor to his family and his country, and died poor. He had many successes, but they were short-lived. He is virtually unknown outside France, and even there few have heard of him. His legacy is limited to three streets—in Paris, Lille, and Lourmarin—and to a sign on a school in Lourmarin. His inventions contributed considerably to the economies of France, Britain, and Poland, and his designs were later used throughout the world.

It is unfair, indeed inaccurate, to dismiss him as an impractical dreamer just because he didn't become rich from his efforts. But his life story raises a number of issues. Why was he taken advantage of so often? Why did he have such dependable connections with members of the nobility, but not with his own government, and not with possible investors? Why didn't he form associations with investors who would have been willing to cut him an honest deal?

One trait that relates to the first question was his political naïveté. He was quite emotional and impulsive, which often led him to take risks in his personal life that seem silly in retrospect—he foolishly defended some unpopular soldiers during the French Revolution and fought a duel over it. Sensing naïveté, unscrupulous people will always move in to use it to their own advantage. He also seemed to have a hard time delegating; while he would welcome assistance from friends, relatives, or experts like the clockmaker, he seemed to want to micromanage every effort. This has a familiar ring to anyone who has been involved in product development. Engineers always want to delay introduction of a product to add more features and perfect it, while the marketing and sales staffs want to get the product out and establish early market share. Managing this tension correctly can be the difference between success and failure for a new product; the danger of introducing a product too soon is that the inadequacies and bugs can ruin a company's reputation, but introducing it too late means that a competitor may get too well-established to dislodge. In this case Philippe was the engineer with no marketing and sales force to balance him out. Otherwise, why didn't he and his brothers set up a business making hydrostatic lamps before his design was copied? Why didn't he attempt to set up a linen-spinning business with an investor immediately when he saw the political situation crumbling? Perhaps he just couldn't get himself to work with an investor, possibly because of his class consciousness, and possibly because he didn't appreciate the important role of investors in setting up a successful business. Also, like many inventors today, he may have had the feeling that once he had a patent, the income should flow rather automatically. It also seems that he had too much faith in the power of a patent to protect his ideas, and this cost him dearly. He should have enlisted the services of a dependable attorney in each country where he lived, but he apparently didn't. Finally, he clearly wanted to spend his time doing what he loved—namely, inventing. When he got in trouble, his approach was to invent something else, which often made his financial problems worse—each new invention required investment to patent and demonstrate it, money he didn't have.

Having said all this, clear societal inadequacies are also clearly at work. The patent laws and inter-country agreements should have been constructed well enough to protect him. The criteria for the million-franc prize should have been better spelled out so that he could have determined more accurately his financial risks before embarking on the venture. The petty bureaucrats should have been intelligent enough to take a longer view of the implications for the French economy than to reject his work to avoid paying the prize—there was no technical justification for rejecting his machines, as he had clearly demonstrated their practicality.

Ideally, there should be a means by which a person of his ability could spend the major part of his life productively. He was already motivated early in life to invent; he did not need financial incentives, and it would be facile to suggest that his financial difficulties sharpened his abilities—the opposite is more likely. Had he been able to devote himself to his passion, who knows what other useful things he might have come up with? Perhaps if the limited liability company had been a reality during his lifetime, he might have set up a successful corporation to build his inventions. A person who fails in one environment may succeed in another.

From the perspective of the five essential conditions, Philippe was in many ways fearless, partly due to his aristocratic upbringing. French society at the time rightly prided itself on its scientific knowledge and expertise and provided him with an excellent scientific education. The generous prize money was an indication of the value placed by the society on innovation, although the stultifying effects of monarchical rule had not yet been replaced by more democratic institutions, so the mechanism for providing meaningful monetary rewards was not well developed. While there were multiple opportunities for him to survive (by living in Austria and Poland), there were not enough good options, at least from his point of view, to get his ideas embodied as salable commercial products.

Considering the misfortunes and setbacks that Philippe de Girard experienced during his lifetime, his inventive output is nothing short of extraordinary. He was truly a great innovator.

Thomas Edison (1847–1931)[10]

Thomas Alva Edison, probably America's most celebrated inventor, was born into a very different world from Galileo Galilei and Philippe de Girard. He was not born into wealth or nobility, although his family would have been considered comfortable. His mother was a schoolteacher, and his father was a businessman. Both had a strong interest in making sure their children were given a classical education. As the youngest of seven, Thomas expected that he would have to make his own way in the world.

He was born in Milan, Ohio, a town near Lake Erie that was experiencing a flourishing grain trade at the time. The proximity of one of the Great Lakes made it possible for farmers to efficiently ship their crops back east through the Erie Canal, and Milan became an important shipping port. His father ran a business making high-quality shingles. While Thomas was a child, however, railroads were built to service the farmers in the area more cheaply, and the town began a decline. Anticipating the loss of income they would suffer, the family moved to Port Huron, Michigan, where his father got into the grain and feed business.

Thomas was considered to be fragile as a child, so he was kept out of public school. Instead, he was taught by his mother at home, which turned out to be a blessing because she made sure he got an excellent education. According to his biographer, he spent only about three months in public school. That was fine with his mother, because she had a low opinion of the quality of the public schools. Under her guidance, he became an avid reader and developed a taste for books and literature. By the age of twelve he had read a number of advanced books such as Gibbon's *Decline and Fall of the Roman Empire*, Thomas Paine's *The Age of Reason*, and the *Dictionary of Sciences*.

Coming from a family with an independent spirit encouraged him to read with a critical eye. He claimed that from an early age he took to questioning the validity of every statement he read in a book. This obsession for detail and rigor was to become a defining characteristic of his personality and his approach to inventing. He developed the irritating habit of asking questions about anything and everything, which, oddly enough, led a number of adults, including his teacher at the school he briefly attended, to believe he wasn't very smart. Apparently they misinterpreted his constant questioning as a lack of understanding. He was considered one of the poorest students, and this same teacher referred to him as "addled." His mother knew better, of course, so she was happy to get him out of that environment.

By the age of eleven, Thomas was already experimenting. He had become interested in the elementary science experiments described in Parker's *School Philosophy*, a textbook in use at the time. He tried to duplicate every experiment he could, a practice that he continued throughout his teenage years. Fascinated by chemistry, he set up a laboratory in the basement of the family house where he carefully arranged several hundred bottles of substances that he had bought at the local drug store. Amusingly, he labeled all the bottles as poison so no one else would take them, and his mother made him put a lock on the door. He wasn't always so careful, though, as he once started a fire that he couldn't control in a barn, and the barn burned to the ground.

Buying chemicals and conducting experiments cost money, so young Thomas decided he needed more money than he was getting as an allowance from his parents. Initially, he and a friend sold vegetables in town from the family farm, which was his introduction to the business world, and it certainly was a step up from the traditional lemonade stand. He tired of this after about a year, however, and looked about for some other way to support his experimental activities.

What happened next would provide Thomas with a profound educational experience that no school or university could ever match: he learned how the world of business in the United States really worked. This knowledge would enable him to thrive and prosper, in contrast to Philippe de Girard, who never acquired the corresponding commercial knowledge and skills in Europe.

What happened was the fortuitous concurrence of several conditions of the times. First, it was a time and a place where a twelve-year-old boy could be out of school and take on a job, not something that could happen in industrial societies today—that is, he was treated very much as an adult. Second, the area he lived in was experiencing major changes—new farms and industries were being built to meet market demands from the large eastern-US population—and the population of southeastern Michigan was growing rapidly. Third, an important local event took place—the extension of the Grand Trunk Railroad through Port Huron to Detroit, already a major city. Fourth, during these years the great Civil War erupted, which created a desperate desire for news from the battlefields.

In looking about for some way to earn some money, Thomas recognized that the new railroad offered an opportunity. The new train route left from Port Huron early in the morning and returned late at night. He could pick up the *Detroit Free Press* newspapers in the afternoon and sell them on the way back, as well as on the following morning. He inquired about the possibility and applied to the railroad for the privilege of doing this. Within a month or so, he got the concession. His mother and father reluctantly allowed him to embark on this adventure, which must have taken incredible restraint on their part, given that three of their children had died. It is difficult to believe that any modern mother would ever consent to a scheme like this, but that was another time.

Soon after Thomas started his new job, he realized that he could augment it by selling produce, butter, and magazines to people on the train, so he went into business with a couple of friends. To help keep the train employees friendly to his pursuits, he gave them discounts. Upon learning that the train had a baggage car with three compartments—one for baggage, one for US mail, and one for smoking—he asked for and received permission to use the smoking compartment for storing his newspapers, periodicals, and food. It turns out that no one else used it because of its lack of ventilation and the availability of a more convenient smoking car on the train.

Thomas had to board the train at 7:30 in the morning, and he didn't get back home until around 9:30 in the evening. This meant that from about 10:00 AM until 7:00 PM he had no demands on his time. He then got permission to use the compartment in the baggage car to install his laboratory equipment. This meant that he didn't have to limit his beloved experimenting activity to the weekends. And experiment he did, spending virtually all his spare money on equipment and chemicals to indulge his passion for replicating every physics and chemistry experiment described in the books he got from the library.

But Thomas didn't stop there. After talking to people on the train, he realized that a lot of local news from the towns along the train route wasn't being covered by the city newspapers and magazines. So he bought a small printing press, installed it on the train, and became the writer, editor, publisher, and distributor of a small weekly newspaper. Eventually he was selling about four hundred copies at three cents each. He sometimes netted twenty dollars or more a month from this enterprise. To put this into perspective, the average monthly wage for adult workers supporting families was in the range of thirty to forty dollars. With all this work, he took on a friend to help in these endeavors, dropping the food and periodical sales efforts.

The events of one momentous day gave Thomas valuable hands-on knowledge of the workings of the market, as he himself attested. The newspaper posted late-breaking news on bulletin boards in its offices, and that day the dramatic news that several tens of thousands of soldiers had died in battle was displayed. Seeing the crowds of people there, Thomas realized that people along the train route would be equally moved and eager for details. He immediately hit on the idea of stimulating newspaper sales by telegraphing the stationmasters along the route to post headlines of the news. He convinced the telegraph operator to contact the stationmasters and ask them to display the headlines on the bulletin boards normally used for displaying arrival and departure times. To sweeten the pot, he gave the operator three months of *Harper's Weekly* and other papers.

On a normal day, Thomas would sell about one hundred papers, but he realized that this was a big, rare opportunity. He thought he could sell a thousand, but he only had enough money in his pocket to buy about three hundred. In a stroke of audacity and desperation, the fifteen-year-old walked into the office of the editor of the *Detroit Free Press*, saying he had important business to discuss. Two men were in the office, and he asked them for credit for the seven hundred newspapers he didn't have cash for. At first the editor refused, but his associate, impressed by the young man, convinced the editor to grant the credit. Thomas and a friend then lugged the thousand copies to the train and prepared to sell them. After two stops where crowds gathered, it became apparent that the advertising campaign was even more successful than he had expected. Recognizing the opportunity, he raised the price from five to ten cents. The papers continued to sell out, and by the time he reached Port Huron, the last copies were sold for twenty-five cents apiece. He made over a hundred dollars that day, but, more important, he learned how to operate in the business world.

Thomas was a voracious reader and spent many afternoons in the Detroit Public Library. He would pick a topic and then read every relevant book on the shelf. His memory for detail was legendary, and he used this talent not only for absorbing and analyzing what he read but also for accurately remembering the methods, equipment, and results of thousands of experiments that he and his associates performed. Later he experimented with his eye movements and trained himself to take in a whole line of text at a glance. He believed this technique was something that should be taught in schools, as it greatly enhanced reading speed.

Encouraged by his growing knowledge of chemistry and mechanical devices, Thomas set up a rudimentary telegraph link with a friend. This experience taught him that for something as complex as a telegraph to be practical, it wasn't enough to develop a single device. It was necessary to think through the entire process, which in the case of the telegraph included a means to create the "dots and dashes," an energy source to transmit the signals, an efficient medium to communicate the signals at a high-enough intensity to be detected at the other end, a means of receiving the signals, and a means for observing and/or recording the dots and dashes. At the time, none of these steps were fully understood or developed. This experience served him well; when he was working on developing an efficient electric light bulb, he recognized that before his invention could really be utilized by the public, he needed to design the entire electrical generation and distribution system. All inventions occur in a societal context, which may or may not be well understood by the inventor. Thomas was an innovator who worked hard to understand this context.

A funny example of this occurred in 1868 when he invented a vote recorder, which he thought would be welcomed by federal and state legislative bodies because it would greatly speed up proceedings by eliminating long, laborious role calls. However, he found it impossible to sell to Congress. The reason? He was told explicitly that such a capability would make filibustering much more difficult; the congressmen were quite happy with the time wasted, thank you very much. After that experience, he decided to only invent things for which he was sure there was a genuine demand.

Since Thomas Edison did have some success in the business world, many people think that he was motivated by profit. This was simply not true. He wanted to get enough income to fund his ambitious inventive activity and live comfortably, but not to make large amounts of money. Like de Girard, he wanted to invent useful and practical things. As a youth, he spent all his income buying laboratory equipment for his experiments, and as he got older he spent most of his income on his inventions. While he understood how supply and demand worked, he was remarkably uninterested in the stock market or in the financial side of his businesses. There are several charming stories about this side of the great man.

One is that for two years Thomas Edison used a primitive system of two hooks—one hook for bills, the other hook for reminders of the money owed to him. Periodically a messenger would come around from the bank with notes and penalties, and he would then settle his accounts. He finally hired a professional accountant to handle the transactions and prepare the books, but he never trusted the balance-sheet numbers until his bills were paid, and he could see the money in the bank-account balance.

Another story concerned some patents for developing improved stock tickers for the Gold & Stock Telegraph Company. The president, General Lefferts, wanted to buy out the rights and patents from him. When the general asked how much he wanted, Edison, who was hoping for $5,000 but would settle for $3,000, asked the general to make an offer. The general then offered $40,000. Edison later said that he almost fainted, but he was able to squeak out that he thought that was fair. He was given a check for that amount, but he didn't have a bank account. He must have mentioned his lack of a bank account to the general; when Edison went to a bank to cash the check, the bank tellers gave him a bag full of small bills, a cute practical joke. After that, Edison got a bank account and deposited the money.

A similar story is about a patent Edison sold for enhancing telephone transmission. He was prepared to hold out for $25,000, but he was offered $100,000. This time he placed a condition on the deal that he be paid in installments of $6,000 each for seventeen years because he recognized that "I would soon spend this money experimenting if I got it all at once, so I fixed it that I couldn't. I saved seventeen years of worry by this stroke." He repeated this scheme later with another telegraph-related invention to get another $12,000 per year.

During his lifetime, Edison's accomplishments were legendary. With the possible exception of the phonograph, his inventions were major improvements in already-existing technologies, such as the telegraph, the telephone, the stock ticker, the light bulb, the dynamo, the carbon microphone, the fluoroscope, and many others. In total, he patented 1,093 inventions.[11]

Apart from his inventions, his innovations included the establishment of his industrial research laboratory in Menlo Park, New Jersey, and his "total systems approach" to invention. Whereas other inventors might focus on simply making an efficient light bulb, Edison recognized that the potential for lighting homes and businesses meant that bulbs had to be cheap and easily manufactured in quantity, a bulb must have a long lifetime, and, most important, there had to be an electrical distribution system that could provide the electrical power necessary and be safely and cheaply installed in homes and businesses. As he and his laboratory staff made each improvement and submitted patents, they were working on the next part of the overall problem.

The reason I devoted so much time here to Thomas Edison's childhood is that I wanted to provide one of the more detailed accounts of the early life of so innovative a man and to show how a gifted young person reacted to the cultural and historical crosscurrents he was caught up in. He was brought up in a household and culture without fear—his parents encouraged his critical thinking and removed him from the influence of a stultifying school. He had access to information—the books his mother gave him were as good as she could provide from the family bookshelves and the public libraries, and later he had access to the library of a big city. His parents provided the essential initial guidance in reading and critical thinking, and later the books provided guidance in the use of information with their descriptions of chemistry and physics experiments. Later in his life, as his inventions proliferated, his patents were sold for enough money to make it possible for him to set up laboratories and continue his inventions and experimentation. The society enabled him to set up services and companies that made him prosperous. He was then heralded and honored for his accomplishments. Finally, the mobility within the society enabled him to choose different types of work, to select the place where he could best build up a profitable company, and to move freely—in short, he had a multiplicity of opportunities in the kind of work he chose, in the place he worked and located his laboratories, and to whom he sold his patents.

Other Great Inventors and Innovators

There have been many other outstanding innovators who might have been selected, and I have mentioned a number of them in previous chapters. I chose Galileo, de Girard, and Edison because collectively they are representative of innovative people, not necessarily because they were the best, the most creative, or the most prolific or because they had the most impact on history. Rosen (2010) would probably claim that the inventors and scientists who developed the steam engine deserve the "most impact" title, and I wouldn't attempt to refute him. Others might well argue that the title of the most prolific innovator should belong to Leonardo. Einstein might well deserve the title of the most imaginative of the white-collar scientific world.

The title of the most persistent inventor might well go to John Harrison (1693–1776), the inventor of the chronometer.[12] Like de Girard, he was motivated by the possibility of a prize, in this case £20,000 (equivalent to millions of dollars today), to be given to the inventor who could develop a practical shipboard instrument or technique that would determine latitude to within half a degree of arc at the equator, or roughly fifty-six kilometers (thirty-five miles). To achieve this accuracy using an onboard clock and time measurements of the highest point of the sun requires a time accuracy of two minutes. While that doesn't sound difficult today, it must be remembered that this accuracy had to be maintained throughout a voyage across a significant part of the globe and involving several weeks of time. For a clock mechanism to work in a shipboard environment, it had to compensate for the effects of the ship's motion, temperature and humidity variations, and even variations in the force of gravity. Over three decades he successfully addressed these issues, finally producing a small clock that maintained an accuracy of five seconds over a voyage of eighty-one days. The famous explorer and cartographer Captain James Cook used one of Harrison's chronometers and had high praise for its dependability and accuracy. Harrison finally received his prize money at the age of eighty, after an unbelievable number of setbacks due to jealous colleagues, competitors, and ignorant bureaucrats.

In reading the biographies of great inventors and innovators, one can't help but be struck by the ups and downs, stops and starts, successes and failures, and the reversals of fortunes they experienced in their careers. It's clear that innovation is just not an orderly process at all. Where lesser individuals would have given up, great innovators persevered. They kept looking for answers for years after their initial efforts did not bear fruit, confident that somehow what they were trying to do was possible. This innovative spirit is one of the most truly noble characteristics of the human race.

16. Innovation and Governments

"Ocracies: autos, plutos, and demos"; monarchies, dictatorships, and communists; how they rank applying the essential five conditions; surprise: democracies win

Humanity has experienced a variety of governmental forms, giving rise to a proliferation of "-ocracies" and "-isms." Like any terms used to describe whole societies, within each category there are many variations, and the categories are not totally distinct. So it's necessary to be careful when attributing traits and making judgments. Each country's government and society have conditions and features that are unique and that don't fit precisely within any general categorization. That said, it is possible to come up with some pretty accurate assessments of how innovative a society was in the past based on which form of government was in power. These assessments are equally valid today.

In this chapter the following types of government will be considered: (1) autocracies, (2) monarchies, (3) plutocracies, (4) dictatorships, (5) communist governments, and (6) democracies. By now it will come as no surprise that autocracies will be assessed as the least innovative governments, and democracies as the most innovative.

Innovation under Autocracies

Most ancient civilizations were governed by a single ruler, king, or emperor. The ruler was supported by a military organization and a bureaucracy that maintained order, taxed the populace, and carried out governmental functions. Such systems of government are called autocracies. The power of the autocrat was far-reaching and varied from absolute ownership of everything and everyone to a preponderant ownership with some power sharing. A good example of power sharing was the ancient Hebrew society, where the religious authorities had some independence from the kings. But the distinctive feature of autocracies was that the societal values were defined by, and embodied in, the office and person of the autocrat.

The primary concerns of the autocrat have always been (1) maintaining control of the people in order to stay in power, (2) expanding his (sometimes her) scope of influence and defending it against outside enemies, and (3) building structures, such as temples, cities, pyramids, tombs, and palaces, that visibly demonstrate his power and serve to intimidate the populace. Typically, monetary support from the royal treasury went to the military to support war activities; to architects and builders to design and erect palaces, monuments, and temples; and to artists and craftsmen to create sculptures, mosaics, paintings, tapestries, and ornaments to decorate the buildings. While the ruler had to have some concern for maintaining enough agricultural production and medical capability to keep the subjects of the realm healthy enough to staff the military, construct buildings, raise the crops, and work the mines, rarely were the common people's needs and desires a major concern.

It is important to recognize that the people generally bought into these values and identified with the autocrats. They celebrated the military victories, they were proud of the buildings and tombs, they appreciated the artistry and craftsmanship of the public sculptures and decorations, and the private lives of the privileged families were the stuff of everyday gossip.

This rather limited set of concerns of autocracies meant that innovations were encouraged, adopted, or tolerated only if they were of value to the autocrat in promoting his control, expanding his sphere of influence, enhancing his structures, or contributing to his displays of power. Thus, technical innovations were usually practical ones—the father of invention dominated, although mother played a part, too. Craftsmen and artisans were allowed and encouraged to develop new techniques for making beautiful furniture, decorative objects, and jewelry. Architects could develop new concepts to make buildings bigger or more impressive. Improved weapons that made the military stronger were encouraged. Better accounting methods and record keeping made projects more manageable. Mathematics and astronomy were used to construct and orient buildings, predict eclipses, to make astrological forecasts, and to time agricultural activities according to the season. Medical knowledge was used to promote the health of the royal family and members of the court. These kinds of ideas could get the approval and support of the autocrat and his bureaucracies, so these were the areas of innovation that autocracies have historically supported. However, because the interests of autocracies were so limited, innovations were severely circumscribed: the payoffs had to be quite obvious and had to accrue to the powerful, while improvements in the lifestyle of the subjects were not supported or considered relevant. In fact, new ideas that might improve people's lives in other ways were often considered subversive and were actively discouraged.

Autocracies have existed since civilization began. The empires of Egypt, Mesopotamia, Persia, Babylon, South America, India, and China were classic examples from the past. Autocracies evolved somewhat from being purely personality cults, in which the ruler had absolute power over the individuals. The dynasties of China, for example, which existed until the last century, were defined by longer eras, and the individual emperor was more constrained culturally to behave within certain limits than the autocrats to the west. Also, the sheer size of the bureaucracy limited the emperor's ability to make fundamental changes, such as a change of religion, by a pharaoh's edict.

What all autocracies did was establish elites. The autocrat needed a loyal group of people that would carry out the functions of government; collect taxes; supervise the foresters, miners, stone workers, construction workers, and farmers; plan new cities and buildings; lead the armies and navies and plan military activities; and manage commercial activity. Also needed were priests to perform religious ceremonies; technical experts to plan and supervise the building of canals, aqueducts, roads, bridges, ships, and vehicles; and astronomers and astrologers to monitor the heavens. These elites were privileged and depended on the goodwill of the autocrats. They received education, money, titles, and homes. They were the source of the white-collar innovations.

The presence of the elites posed a barrier to innovation because they were usually given monopolies in commercial concerns within the economy. If one family was given rights to the chariot trade, for example, it did not take kindly to competition from an inventor of better chariots. Moreover, that family was not highly motivated to spend money on improvements or to develop cheaper chariots, since it already had a lock on the market.

From the autocrat's point of view, this made perfect sense. The ruler needed trustworthy people, and trust was far more important than competence. The interconnections of the elite families ensured loyalty to the autocrat and to the system. They jockeyed for the autocrat's favor and competed, in some sense, for the granted monopolies, but they acted collectively as a class. People outside the inner circles who petitioned the kings with a novel proposal risked having it either rejected or stolen by an insider. Innovation can hardly thrive under such conditions.

The idea that cronyism and favoritism might be considered corrupt is a relatively recent concept. It is important to realize that historically these practices were considered business as usual and not at all unethical. In Roman society, bribery was a widely accepted practice. Nepotism, patronage, and giving monopoly rights to court favorites have been time-honored practices by autocrats and monarchs. These practices were considered unjust only by the ones who lost out to someone else. It is only with hindsight that we can look back and realize that the loss of the efforts of smart people and the suppression of innovation held the societies back from building on the accumulation of inventions. Only where the concentration of power broke down could the concept "the invention of invention" (Landes) or "technology begets technology" (Diamond) develop.

The problem with monopolies within autocracies is a specific example of a larger principle: *any significant change to the status quo will "gore someone's ox.*" In other words, anyone proposing an innovative idea will run afoul of someone else's established turf. If that person is in a powerful-enough position to quash the idea, he will almost certainly do so. In any form of government where power is concentrated in the hands of a few, there will be a strong likelihood that any new economically significant design, machine, or technique will threaten someone's established commercial or financial position, and that person will oppose it. That was true in the past, and it is equally true today. It explains why autocracies didn't develop their technologies further: at every step there was someone in a position to stop or at least severely retard the use or evolution of any new invention. New inventions were treated as isolated events. They did not have environments encouraging to sustained and self-reinforcing inventive activity.

In applying the five conditions for innovation, it is evident that autocracies by their very nature are, at best, discouraging to inventive activity, and usually they have been downright hostile. Since the autocrats have largely determined the attitudes of the society at large, these societies have usually been hostile to most types of innovative activity.

Fear has traditionally been important to the autocrat for maintaining control over his subjects. Even within the inner circle of noble families, the fear of losing the approval of the autocrat was ever-present, so risk taking of any kind was discouraged. For the subjects outside the inner circle, the fear of running afoul of the autocrat or his court, their inability to become part of the inner circle, and the daily reminders of their low status all conspired to discourage any significant initiatives. On the other hand, in the places where the subjects worked the operators would have been more encouraging to workers with new ideas, as the operators would reap the benefits, so fear would have been reduced in these circumstances.

The limited access to information has been less of a factor, as was guidance in its use, but autocrats have historically limited information access to trusted associates. This has meant that only a small percentage of the population was in a position to make significant contributions. The subjects outside the inner circle were typically illiterate, so they only had access to information known to their elders and associates. In such an environment, innovations that benefited the populace were limited to small inventions like new tools and techniques.

Where autocracies have failed miserably is in providing a multiplicity of opportunities to their citizens. Since autocrats believed that strong leaders like themselves were essential to good governance, they unconsciously or consciously subscribed to the notion that economic life should be run in a monopolistic fashion, ultimately under their control. This attitude pervaded the system and discouraged competition in all walks of life. Instead of ordinary citizens deciding whether new ideas or inventions were worthwhile, only the opinion of the autocrat and his favored few mattered.

Innovation under Monarchies

After the breakup of the Roman Empire, small kingdoms eventually developed across Europe, and each king aspired to head an autocracy in the old model—after all, it was the primary governing model to imitate. It is true that it was not the only model, but the alternative, ancient Greece, was considered a relatively unimportant and short-lived experiment; plus, there were no kings to serve as examples. The memory of things Greek was reduced to honoring that society's great thinkers, and the Greek experiments in democracy were de-emphasized. Fortunately for the world, three conditions kept the monarchs from establishing autocracies. First, no kings or queens were able to extend their control over a significant part of Europe; this condition limited the scope of any single ruler's wealth and power. Second, the new kingdoms evolved from an existing power structure of feudal lords and their descendants, so the rulers had to negotiate with them in order to govern successfully; the aristocracy was not willing to cede total power to the kings. Third, the religious life of Europe was dominated by the Catholic Church, which was not about to abrogate its spiritual authority by allowing demigod status to earthly rulers, although it supported the watered-down divine right of kings until relatively recently. These conditions reduced and eventually removed some of the most powerful tools the autocrats had to control and intimidate the populace. The result was a set of kings who did not succeed in becoming autocrats in the classical sense. They weren't demigods, they didn't always own all the land, they didn't own the populace, and their attempts to acquire land, control the populace, and to fight foreign wars ran into the wall of the competing interests of the aristocracy that funded them. In spite of these limitations, however, they still had enormous power over the lives of the people.

Reading the history of the monarchies during the Middle Ages is like reading a bad soap opera: it is a never-ending tale of power struggles and shifting alliances between bishops and kings, lords and kings, bishops and landowners, kings and kings, lords and lords, and sometimes even between bishops and bishops—a litany of plots, intrigues, assassinations, wars, and political marriages. Looking back on this today, it looks farcical—costly and deadly games played by the upper classes, all vying for power and seeking to reestablish the old autocratic ways. Kingdoms were created and destroyed, and their boundaries shifted back and forth. Imperial ambitions within Europe were realized only by the Holy Roman Empire, which even at its largest did not dominate the region. None of the monarchs managed to dominate Europe, which remained fragmented. For this we should all be grateful.

The old autocratic ways would not return to Europe. While the elites were fighting it out, changes were taking place. Trade increased by sea between Europe and other societies and by land between its kingdoms. A merchant class developed outside the monopolies and favoritism of the monarchies, and many became quite wealthy. While the aristocracy sniffed at the nouveaux riches and resisted their attempts to gain political power, many traders and merchants were accepted into the upper classes. Some made it in through convenient marriages, while others bought the titles of nobility and took over the lands of bankrupt titleholders.

The monarchs of Europe found they could derive revenues for the royal coffers by granting monopoly rights to favored people. In England in 1621 there were about seven hundred such monopolies[1]. Competition became more common as anti-monopoly measures were put in place.

The monarchs, particularly those in western Europe, became weaker as town officials, landowners, and merchants became politically stronger. These classes continually demanded more rights, rights that eventually would apply to almost all citizens. As this happened, more people became literate and better informed; fear of the monarchs declined, and innovative activity increased. The benefits of inventive activity became more widely recognized, and various power groups began to act in an organized fashion around national interests. They began to recognize that innovative activity increased the power of the nation and improved its military capability, and that the national production and export of goods could be a source of wealth for all the elite. Innovation became accepted as a path to prosperity, and with it came an acceptance of competition, to some extent. Most monarchies became more plutocratic, and eventually they became more democratic. Europe still has monarchs, but they have become national rallying symbols, rather than power centers, and the governments are democratic.

Today the Middle East has some kings who have more than symbolic power, but the countries are better described as plutocracies; the king of Saudi Arabia, for example, shares power with the large Saud family. The classical monarchy is now rare.

Innovation under Plutocracies

In this book a plutocracy is defined as a societal form where a few powerful people share power and exercise control over the rest of the society. The ruling classes may directly govern or may exercise control through proxies and agents. This definition is in line with the conventional one, which identifies plutocrats with wealth; but wealth and power go together. A plutocracy differs from an autocracy in that power is shared, and the power structure attempts, not always successfully, to prevent absolute power being seized by a single person.

All countries have elements of plutocracies; there are always powerful people who exert a disproportionate influence on a country's politics and economy. As we have seen, monarchies had to negotiate first with the aristocracy and church officials, then with merchants, then with city and town officials. As the monarchies became weaker, the societies became more plutocratic, with multiple power centers. Dictatorships, while usually based on military power, require a power base of rich people; hence, they have elements of plutocracies. At the other end of the spectrum, many so-called democracies are not democracies at all, but are more accurately described as plutocracies. While they may have a formal election process, the choices are actually determined by one or more powerful groups that act in their own interests. Whether these groups actually hold office or act behinds the scenes isn't the distinguishing factor; rather it is the extent to which these groups control the leaders, the legislatures, the courts, and the economy.

Plutocracies also differ to the extent they act on behalf of the public; some are more predatory, in that the economic institutions are designed to funnel most of the society's revenues into the pockets of the plutocrats, while some are more public-spirited and benign, in that they accept the idea that a healthy and prosperous society benefits them proportionately (note that they may not be benign in other respects). Thurow's distinction between an "oligarchy" and an "establishment" graphically depicts these two types of plutocracies: an establishment believes it has a stake in the community and is willing to make short-term sacrifices to ensure that the rest of society succeeds, while an oligarchy doesn't identify with others in society and doesn't care if the majority of people struggle.[2] Establishment plutocrats are often philanthropic and spend their wealth to improve their societies, while oligarchic plutocrats typically put their money in foreign bank accounts, ready to move out if things get too hot at home. Sometimes people can be both at different times of their lives: human nature being as complex as it is, there are several examples of oligarchs who became establishment plutocrats in their old age. After focusing on achieving money and power for most of their lives, they turned their energies later toward philanthropic and educational endeavors.

The Venetian Republic was a plutocracy that deliberately limited the power of its elected rulers. Its electorate was the Great Council, initially composed of about five hundred member families that determined who would be in positions of power. Venice based its prosperity on trade rather than direct control by conquest. By focusing on mutually beneficial trade, this relatively small city-state was able to survive and prosper for several hundred years. It allowed non-noble merchants to become wealthy and marry into the power structure, the few hundred families that ruled the republic. The economy's dependence on the efforts of craftsmen, sailors, and workers supporting the businesses, as well as the close proximity of everyone, resulted in a society friendlier to the lower classes. Thinkers, scholars, and artists were held in high regard, as we saw with Galileo. These conditions encouraged creativity of all kinds. Venice made improvements in the size and maneuverability of ships, developed navigation techniques to take advantage of the newly invented shipboard compass, and formulated improved accounting techniques. Other nearby city-states developed near-plutocracies headed by one noble or merchant family, or by a small group of such families, and benefited from the trade around the Mediterranean and beyond. They, too, supported some innovative activity. Filippo Brunelleschi's dome atop the Florence Cathedral is an example of an impressive innovation of the times. Unable to use conventional techniques because of the size of the area to be spanned, Brunelleschi developed a two-shell technique that made it possible for workmen to gradually construct the dome from the rim up. In the process he invented a special hoist for raising the millions of bricks to where they could be laid in place.

The Italian city-states didn't further develop a middle class. Except for supporting a few outstanding individual scientists and engineers, they didn't participate significantly in the explosion of innovation that took place in northern Europe. In part, this appears to be due to the fragmentation of the region. As was shown earlier, size matters when it comes to innovations with commercial implications. Developing a commercial invention costs money as well as time, so the potential market has to be large enough to justify the initial investment. City-states typically had populations of a million people or less, so products that would only sell locally could not require a significant investment. By contrast, France and Britain were real nations with populations in the tens of millions, so inventions like steam engines that required a lot of upfront money and time could be justified. Italy did not become a nation-state until 1870, so there was no national government or institutional framework to promote industrialization or patents, set up nationwide libraries, or reward inventors or entrepreneurs. The ruling families of the city-states did not have a strong stake in educating the general public or in promoting a competitive environment, so they didn't make serious efforts to unify.

There was another effect at work in northwestern Europe—namely, a multiplicity of influential power centers that included the heads of cities and towns. The mayors of these local governments had nothing like the power of the Medicis. The various plutocratic groups had different interests, and the conflicts kept any one group from taking control. This situation led to compromises, such as the establishment of the House of Commons in England, originally made up of local leaders, initially appointed, but evolving to elected politicians. By contrast, in the Italian city-states the merchant classes became absorbed into the aristocracy, so there was no multiplicity of powerful groups; frequently two or three feuding families vied for control. Plutocracies that are singular are prone to takeover by dictatorships, and frequently are predatory. Plutocracies with multiple power centers, on the other hand, are more likely to be benign, since competing interests will collectively better represent the interests of the populace; they are the ones that typically have evolved into democracies.

It is significant that the initial expansion of innovation in the 1700s took place in the multicenter plutocracies in Europe that had not yet made the transition to democratic, elected governments. The electorate at the time was limited to relatively few male property owners. This shows that significant innovation can occur in societies that are not fully democratic. It can happen if the plutocrats support innovative activity. The continued explosion of innovation of the 1800s and 1900s, however, did occur primarily in democratic countries like those in North America and Europe.

It is also true that plutocracies can be the engines of growth and prosperity, if the plutocrats are motivated more by national interests than by individual greed. They can use their wealth to promote rapid industrialization and to incorporate the latest technologies. Prime examples are Japan, South Korea, Taiwan, and, most recently and dramatically, China. Not only does the impact of establishment plutocrats raise the country's GDP, but it also supports democratizing forces and makes the latest technologies accessible to its innovative citizens, who will adapt them, improve them, and use them as springboards to new inventions.

Because there is such variation in the way plutocracies operate, it is not possible to come up with a single evaluation in terms of our five conditions. The oligarchic, predatory plutocracies closely resemble the autocracies, while the benign, establishment plutocracies score higher in each category. In establishment plutocracies people are less fearful, there are fewer limits on access to information and education, and innovative activity is more highly valued. However, because societal control is still a paramount concern, they don't do as well as democracies in these areas. As for multiplicity of opportunities, plutocracies do quite poorly: citizens have fewer choices in education and work, the press is more tightly controlled, there are no significant opposition parties, there are generally fewer property rights, there is more corruption, and competition is limited, all of which reduce the opportunities for innovation.

Innovation under Dictatorships

Dictators are modern versions of autocrats, in that their personal power is huge. Dictators rarely step down voluntarily. Unlike autocrats, however, they lack complete control over the populace. To some extent, dictatorships are predatory plutocracies in which the institutional constraints are too weak to prevent one person from taking over, and that person is usually backed by a military force. South American countries have had numerous examples of such dictatorships, a result of centuries of the Spanish and Portuguese legacy of awarding huge land grants to favored families, who essentially treated the native South Americans as slaves and chattel. This policy made it virtually impossible for less fortunate people to own significant property. Literacy was not widespread until recently. Most of the land, buildings, and businesses were, and still are, controlled by a few privileged families. Strongmen, usually military leaders, were able to put together sufficiently large armies to establish themselves as dictators, garnering enough support from a few families to maintain control. They rarely ruled countries for more than a few decades because other powerful families that were left out of the action supported rebellions. As the faces of the dictators changed, so did the people who filled the legislatures and courts; therefore, permanent, independent societal institutions rarely got established.

None of the conditions for supporting innovation are present to any large degree in dictatorships, and competition is virtually nonexistent except for small businesses. Large businesses are controlled by the dictator and his favored cronies. It's no surprise, then, that innovative activity in these societies has been minimal. Most technologies, for example those for drilling oil or mining, are imported, and inventive local people are able to do their magic only in foreign countries where the conditions are better.

Innovation under Communist Governments

The idea of communism as an economic system came about as a reaction to the dismal plight of the laborers who worked at the factories and mines during the Industrial Revolution. Karl Marx and his colleagues developed the idea of communism as a system where the workers would take over the means of production, and thus the system would distribute income more equitably than the capitalism of the times. The Marxists saw private-property ownership as one cause of the problems, and they demonized the capitalists, but they shared the mercantilist doctrine of maximizing the economic production of the nation. They sought to raise the status of the workers and reduce the inequality of wealth and income.

In the twentieth century, communist governments were established in several countries in the world, first in imperial Russia and decades later in China. Just as for autocracies, control was the driving concept, with the hope that this time it would be used not for the benefit of some king or czar and his favored families, but for the general population. To establish the Soviet Union as an industrial nation required a massive change in what was basically an agrarian society, and Joseph Stalin used brutal techniques to make this change: people were forcibly moved from the farms to support the factories, and individual freedoms were suppressed to accomplish the "greater good." In the process, private-property ownership was abolished.

The centralized control of the economy resulted in "demand" being dictated from the top by a series of five-year plans, so the economy was geared toward supplying this demand. The emphasis was on resource exploitation and on the production of apartments, factories, transportation needs, military equipment, and other things designed to modernize the country, bring prosperity, and show off the power of the empire. Indeed, from 1928 to 1973 the annual GDP per capita of the Soviet Union rose from GK$1400 to GK$6,000, quite an improvement from the time of the czars. For many years the ever-growing production statistics seemed to support this approach. However, comparing production statistics to those of the rest of the world overstates this progress. In the Western world, the products reflected the wants and desires of the people for a variety of clothing, food, housing, appliances, furniture, telephones, and entertainment devices, while in the communist countries, the demand was driven by a desire by leaders to maximize the production numbers. The father of invention dominated the mother. As a consequence, the variety and quality of manufactured products was low by comparison.

The highly centralized control inherent in communist states had several consequences that made the system less innovative. First, it made the system highly vulnerable to takeover by strong-arm dictators, which happened with Stalin in the Soviet Union and Mao Zedong in China. Second, all executive, legislative, and judicial institutions were controlled by the same people, so the temptations for corruption and suppression of dissent were too strong to be resisted, as it has typically been historically. Third, like the ancient autocracies, the economies were a collection of state monopolies, with the associated inhibitions on innovative activity. There was no way an individual could invent new widgets or services and then set up a shop to make and sell them for profit; the state controlled all production.

The Soviet government placed a high value on universal education, literacy, and technology. Schools, colleges, and universities were established everywhere. Hundreds of thousands of professionals entered industry, agriculture, education, medicine, music, art, theater, and movies. The creation of this educational infrastructure was probably one of the greatest achievements of the government. Bright prospective engineers and scientists from across the country were able to compete on the basis of merit for positions in the government and at universities. The Soviet Union succeeded in putting the first person into space, Yuri Gagarin, which shocked many in the West. China, too, aggressively promoted education and literacy. So for the innovation factors of access to information and guidance in its use, communist governments get high marks compared to the previous governments, which limited these to elites.

There were societal rewards for innovation for scientists and engineers who served the interests of the state, like the priesthoods of ancient times. They were elevated to the status of heroes in the state-controlled press. However, for blue-collar innovators the story was mixed. In principle, there were rewards, but because the communist dogma made it difficult to give much extra money to people who proposed good ideas, the rewards took the form of certificates, ceremonies, orders and medals, and, sometimes, preferential housing or other nonmonetary rewards. Patents, called "certificates of authorship," were often awarded with money, but it was not possible for a worker with a good idea to reap benefits by getting a percentage of the sales. In attempting to keep workers motivated, the government tended to overuse the symbolic rewards, and this reduced their meaningfulness to the recipients.

Communist states failed to meet the criteria I have set up regarding the conditions of minimal fear and high multiplicity. The brutal treatment of the people by Stalin and Mao to attempt rapid industrialization, the purging of people who might resist or oppose the measures taken, and the top-down organization of virtually every institution created an atmosphere of fear. Better to keep quiet and do what you're told than to risk a suggestion being taken as a criticism. As for a multiplicity of opportunities, options were few; competition was limited. There was no way to develop an invention and market it independently. As I pointed out earlier, all five factors need to be present to have a truly innovative society, and the communist system was missing two of them.

After Stalin and Mao died, the governments of the Soviet Union and China were no longer dictatorships, but rather evolved into plutocracies under the control of their central committees. The new plutocrats recognized serious flaws in their economic models and made attempts to reduce fear and encourage new ideas; these efforts had limited success because an atmosphere of fear can't be eliminated overnight. In their travels it was obvious to them that the Western and Japanese economies were performing much better, supplying their people with a variety of high-quality goods, while the communist societies were having difficulty raising enough food to feed their own people. Many people in communist countries had become aware of these differences from their travels, from radio and television broadcasts, and from magazines and books that got into the countries. The bland assurances of the leaders were not convincing, and something had to be done. In the case of the Soviet Union, it was disbanded in the late 1980s, and Russia and the satellite countries took over a decade to recover from the effects of this action and to establish more market-based economies. The GDP per capita of each country dropped to about half of its previous value in five years. At this time most have recovered, and their economies are in a positive-growth mode, as their governments have established more market-oriented economies.

After Mao, the Chinese instituted major changes in their economy that had far-reaching consequences; this will be discussed further in the next chapter.

Innovation under Democracies

The term "democracies" as used in this book refers to countries with governments that (1) are elected by popular vote, (2) have checks and balances between relatively independent legislative, executive, and judicial branches, (3) have bodies of law that protect citizens against excesses of government or powerful individuals, (4) have laws and enforcement mechanisms that discourage or prevent corrupt business tactics, and (5) have relatively free-market economies[3] and internal competition. There are many countries that claim to be democracies that don't meet these criteria in practice.

People who live in democratic societies are the most fear-free people in the world. In large part this is because the institutions make it more difficult for powerful people to intimidate others. Innovators don't have to worry that they will be hounded for their ideas. Innovation and change have become widely accepted throughout the democratic world as essential to economic well-being. New ideas are encouraged in all areas of life. Religious and class inhibitions are not nearly as strong as they once were, or as they still are in some nondemocratic societies.

The democratizing innovations in transportation, communication, and entertainment have resulted in a world where the advantages of democracy are there for everyone to see. Today only in a few places, like

North Korea and some remote countries, have the leaders been able to prevent their countrymen from having contact with the outside world. As a consequence, the democratic ideals of government-protected individual freedoms, governmental checks and balances, corruption-free institutions and businesses, and universally available opportunities for education and social mobility are now widely accepted values. The fact that the societies that best support these ideals are also the most successful economically is evident to everyone. Leaders opposing democratic ideals are on the defensive and now must pretend they support them even as they subvert them in practice.

Democratic societies have adopted business models that consider patronage, bribes, favoritism, and nepotism to be undesirable. Such actions are described in terms like "corrupt" and "unethical." In these societies it is considered undesirable to give jobs to friends or family instead of to people more qualified by their training, intelligence, and experience; it is frowned upon for a government agency to give out contracts to companies headed by friends or associates, a practice called cronyism. Indeed, capitalism owes its success in large part to the reduction of these practices. Jane Jacobs characterizes this set of attitudes as a new "moral syndrome" that values honesty, respect for contracts, collaboration, and enterprise. She then contrasts that set with the morality of families, tribes, and governments that values loyalty, respect for hierarchy, and obedience.[4] In societies large and democratic enough to support many centers of economic power, it has proved more profitable in the long run to hire highly qualified strangers and to do business with companies on the basis of high performance than to deal only with friends and relatives. Democratic societies are distinguished by having many centers of power and a distaste for monopolies, which enables them to support innovative activity even when competition results in the destruction of a company, or even an entire industry.

This does not mean that the impulse to suppress inventions is nonexistent in democratic societies—far from it. Companies that are threatened by new ideas will use every means possible to suppress them, usually by filing lawsuits or by appealing to legislators. However, they are usually not successful; at most, they slow the adoption of the innovations. This is not true in highly centralized societies, where the people in power can be convinced or bribed to totally suppress new ideas or businesses.

Potential inventors are more motivated when they have protected means of profiting from their inventions, such as patents and the right to sell the resulting products. Democracies have developed institutions and laws that do this. In other types of societies, these institutions and laws are often corrupted or undependable, or inventors can only get their ideas into practice by appealing to the privileged few. Commercial companies in all types of societies are highly motivated to file patents and keep a portfolio of them handy. If its patents are valuable, a company can enjoy a temporary monopoly and gain market share or even dominate the market. During that time it can also license the technique to other firms. It's also a good defensive strategy to have a portfolio of patents because, in case of a dispute over some product or feature, patents can be used as bargaining chips. Thus, most companies in democracies encourage their employees to develop ideas that can be patented by the company, and they reward their employees with bonuses, raises, promotions, and awards.

Democracies have recognized that it is in the best interests of society to promote competition and to prevent excessive concentration of power in business and government. They have antimonopoly laws in key industries and regulate industries (like utilities) that are "natural" monopolies (i.e., where overlapping and duplicative infrastructures that would be required in a competitive environment are not feasible). Similarly, political constraints are placed on governments in the societies' constitutions and laws to make sure the governments don't overly intrude into personal or commercial freedoms. Extensive judicial structures exist to resolve disputes and enforce competitive practices. The result is a great deal of competition in businesses, which creates a very favorable environment for new inventions and ideas. It's difficult for a person or company to suppress innovative ideas, which can find any number of proponents. The multiplicity of opportunities is a key feature of democracies; it is not as extensive in other types of societies. It is reflected not only in the commercial world but in every phase of life.

All democracies have some form of a market economy that decentralizes most economic decision-making and supports competition. This environment encourages new ideas and risk-taking and enables rapid adoption of new good ideas and products; it also enables the rapid rejection of bad ideas and products. John Kay sums up this process very elegantly: "If the strengths of the market economy were encapsulated in a single phrase, that phrase would be *disciplined pluralism*. Disciplined pluralism is the process of perpetual experiment in market economies, in which most experiments fail and are terminated, but the few that succeed are quickly imitated."[5]

In summary, democratic societies score at the highest level in all five conditions supportive of innovation. Citizens are subjected to the fewest fear-producing tactics, information is widely available, education is highly valued, numerous rewards exist for innovative activity, and the support of competition and the limits placed on monopolistic behavior ensure a healthy multiplicity of opportunities.

17. Innovation in Today's Societies

Some countries get high grades, some get improving grades, and others get low grades; size matters, again; why natural resources are becoming less important

It is possible to use the five essential conditions for innovation described in chapters 10–14 to assess a country or even a company or other organization. This chapter briefly describes and assesses a few representative countries. It is clear that in innovative societies the five conditions I have developed are strongly present, while in less innovative societies they are less present or even absent.

Western European Countries

Western European countries are democracies that all provide the five essential conditions for innovation. They have many innovative citizens and, by no coincidence, are prosperous economically, although one might not think so when listening to the economic news today. They all went through the transitions from monarchies to plutocracies to democracies over the past several hundred years. While the details were unique to each country's history, the broad outlines of the transitions were similar:

- Merchants became more influential as trade spread from southern to northwestern Europe. As the population grew, merchants became more numerous and employed many people in the towns and villages. As they prospered, they became more politically powerful, as did the guilds.
- Medieval villages became towns, attracting people from the feudal estates and providing new centers of power. The local officials were originally agents of the feudal lords or monarchs, but over time they gained independence as the towns became new local centers of power.
- Monarchs, who already had to share power with the landed aristocracy and the Church, increasingly lost power to the merchants and town fathers as well.
- A growing sense of nationalism provided the framework of a plutocratic alliance between the different power centers. This led to mercantilism, with its notion that a nation should strive to

produce and sell as many goods as possible. It also set the stage for the Industrial Revolution.

- The Industrial Revolution introduced democratizing innovations that helped to develop a growing middle class and to propagate the recognition that democracy, with its empowerment of the populace, was a viable form of government.

- The ruling classes began to accept the idea that the well-being of common people was in their interest as well, and social safety nets were established.

- Over time, elections were established for parliaments and legislatures with increasing commoner representation and participation.

All this brought prosperity, and the experiences of each country were quickly felt by its neighbors. Led by Great Britain, this trend continued. Although they were held back by their aristocratic history, democratic countries emerged where people had extensive rights and freedoms, including property ownership. Two destructive world wars in the twentieth century threatened to destroy this progress; fortunately the societies not only survived but were able to rebuild in such a way that the five conditions essential to innovation were enhanced. Democratic institutions that reduced fear, expanded access to information and education, and established competitive economies were strengthened. The population recognized the benefits of innovation and supported efforts to use technological advances for peaceful applications.

The countries of Europe have smaller populations than the United States, so the multiplicity of opportunities has historically not been as great. Each country could only support a few top educational institutions, and the variety of industries benefiting from innovation was not as wide. This began to change with the emergence of the European Union, which increased the access of its citizens to universities and encouraged cross-border industries. Aided by the increased emphasis on multilingual capability, the multiplicity of opportunities for innovative people has increased. The countries of western Europe now have the highest economic mobility and educational standards in the world, which have helped sustain the high level of innovative activity there.

The United States

All five essential conditions for innovation are provided at a high level in the United States, and they have been present for a longer period of time than anywhere else in the world. As a consequence, the volume of innovation has been huge over the past hundred and fifty years. While many other countries are equally free from fear, offer similar access to information and education, and provide societal rewards for innovation, the United States offers a higher level of multiplicity, in part because of the sheer size of the country. Major universities in most of the states provide a multiplicity of opportunities for bright students from all over the country. The variety of industries offers many opportunities for inventive minds. The monetary rewards that can be obtained by entrepreneurs have been strong motivating forces and have attracted ambitious people from all over the world. As hard as it is to establish, operate, and grow a business in the United States, it's even harder in most other places, where taxes are higher and governmental limitations on employment and types of operation are stricter. Reflecting their society's values, people in the United States have traditionally had a higher opinion of entrepreneurs than elsewhere.

When it comes to innovation, history has favored the United States in a number of ways:

- In spite of the colonial status of America before statehood, the British rulers were not able to impose their economic model whereby most of the land was owned by the aristocracy. This is vividly described by Acemoglu and Robinson.[1] According to their account, the British attempted to copy the mold of the Spanish and Portuguese in South and Central America. However, the attempt was not successful. The Native Americans were made up of many small tribes, and the population density was much lower than in South America; as a consequence, it was not possible for a few soldiers to round them up and put them to work in any significant numbers. The British government then tried to force emigrants to work the fields, and it attempted to enforce a rigid class structure that would have relegated emigrants and their heirs to perpetual servitude to a few elite landowners. Fortunately, it failed miserably. The settlers successfully resisted, eventually demanding the right to own property and to create elected assemblies. Thirteen colonies emerged, each with a governor and an assembly elected by property owners. These colonies became the first states of the country after independence. By comparison with people elsewhere in the world, property owners in the new United States were relatively free from fear of the colonial authorities, and the expanding federation greatly enhanced mobility, conditions which set the stage for the high level of innovation in the next two centuries.

- Unlike the European countries, the United States was not encumbered with a history of concentrated land ownership and royal monopolies. There were no titled noblemen going back centuries who demanded obeisance from the rest of the population. Social and economic mobility was probably higher in the thirteen colonies in the 1700s (and in the country in the 1800s) than anywhere else in the world. The early settlers brought many crafts and skills from the Old World that helped them cope with their new situation. Later, they were able to adopt and adapt ideas and technologies from Europe—especially from Britain, with whom they had a common language. They recognized that education was important to the welfare and growth of the nation and that they could bring to the average person the information and guidance that were available only to the elites in most of the world.
- The new states were able to establish institutions that copied many of the best features of Great Britain and avoided many of the worst. They established the familiar three branches of government in both the central and state governments; their built-in checks and balances prevented a dangerous concentration of power and reduced the probability of a strong-arm dictator taking over. The existence of many states with independent power also militated against dictatorial or monarchic ambitions. Gradually the central government took on more powers and succeeded in standardizing the currency and promoting commercial, property, and patent laws that applied across the country. These laws enabled inventions to be protected throughout the land and provided a countrywide market for innovations. This dynamic proved to be self-sustaining and made it possible for the United States to reap not only the benefits of size but also the advantages of multiplicity of opportunity and the strong checks afforded by the institutional structures against excessive concentration of power.

The United States established a great number of major universities and research institutions that made it the envy of the world. It established a patent office that eventually processed more patents than anywhere else in the world. It quickly followed Great Britain in establishing laws supporting the limited liability company in the late 1800s, and within decades there were competing shareholder-supported companies across the country. All these developments supported inventions and promoted prosperity, and they promoted an atmosphere where innovators and entrepreneurs were highly valued.

China

China is a fascinating case study because it is currently in the middle of a transitional period in its history that could ultimately result in either of two different outcomes: it could become increasingly democratic, with the populace able to form political parties to offset the power of the communist party plutocracy; or it could remain a plutocracy[2] where the population could reach a historically high, but limited, standard of living. These are not the only possibilities, of course, because theoretically a dictator like Mao might take over the government again, or the plutocracy might somehow take actions that would enable its citizens to have the highest incomes in the world. Most Western observers consider these as highly unlikely, but not impossible, and we could be surprised.

China was autocratic throughout much of its four-thousand-year history, even while its boundaries and rulers changed in response to warlord feuds, peasant uprisings, and attacks from enemies at its borders. In part because of its geography, its governments were generally able to maintain control of its considerable area and to establish a very stable form of government, supported by an extensive civil service that drew on talent from throughout the land. In this regard, the autocracy of China became very different from other autocracies because of the emphasis it placed on merit. By the year 1000, nationwide examinations were held, and those who excelled were given positions in the official bureaucracy. While these examinations were theoretically available to any male, in practice only about ten percent of the male population could realistically participate;[3] still, even ten percent was a lot of people, and it appears that the bureaucracy made sure the examinations were fairly administered. (It should be pointed out that while the system did succeed in getting bright people into the bureaucracy, the knowledge and values embodied in the examinations were philosophical, not practical, a situation which was hostile to innovation. This changed in the 1980s, when many of the members of the central committee were technically trained.) The emperors had many cultural constraints that the autocracies to the west did not, so while emperors changed, the governmental structure remained stable. As was noted earlier, over its history China produced a large number of blue-collar innovations, but most of them were not of the democratizing types and were not accompanied by long-term changes in the society. While major improvements in agricultural methods were developed, they were eventually matched by a comparable growth in population. Economic growth did not exceed the population growth, so the GDP per capita remained near its historic level of about GK$500 to GK$600 well into the twentieth century.

The autocratic form of government officially ended in 1912 with the overthrow of the Qing dynasty and the establishment of the Republic of China, which actually was fragmented into regions controlled by warlords. From attempts to unite the country, two powerful leaders emerged, Mao Zedong and Chiang Kai-shek, whose forces fought for control. By 1949 Mao's forces had taken over the mainland and called it the People's Republic of China, and Chiang's forces controlled only Taiwan.

After Mao's forces took over the mainland, they established a communist government that tried to balance agricultural and industrial production. Initially, land was taken from the large landowners and doled out to the peasantry, but soon most agriculture was collectivized, as in the Soviet Union, with similar disastrous results. The Great Leap Forward (1958–60) and the Cultural Revolution (1966–76) were economic failures and resulted in millions of deaths and widespread suffering. Overall, the party did succeed in raising the GDP per capita to about GK$850 by 1976, although there were serious problems.

After Mao's death in 1976, the government changed from a dictatorship to a plutocracy. The central party exerted its power, and subsequent leaders had to deal with its members. None of the leaders succeeded in getting enough military and political backing to establish a dictatorship. A new constitution, put in place in 1982, included protection of some kinds of private property and elements of free speech. The new plutocracy under the leadership of Deng Xiaoping began to introduce economic reforms that included considerable decentralization, a private sector, and the acceptance of foreign investment and businesses.

These reforms radically transformed the Chinese economy. China progressed from being a Third World–country to being a world power today, both politically and economically. The GDP per capita today is about GK$6,400, almost an eight-fold increase in three decades. China, with a population approaching 1.5 billion people, now has the second-largest economy in the world, and its economic growth continues to exceed its population growth by about 8 percent. This is an impressive record. If this growth were maintained for another twenty years, Chinese citizens would have incomes comparable to those of Europeans today.

China has made great strides in increasing access to the technical information generated by other countries in the past, and this has greatly aided its economic growth. It has built up its libraries and expanded access to the internet. It now sends more students abroad for advanced education than any other country. A large number of these students return home with their newfound knowledge. Because most of the previous scientific and technological knowledge was written in other languages, access has historically been limited to an educated few. The current government is now encouraging proficiency in English and other languages, and it spends considerable resources translating this knowledge into Chinese, using a streamlined alphabet more suited to computers. It is likely that Chinese society will soon have access to technical information on a par with the West. Access to political information is restricted, but how much this will inhibit innovative activity is not clear.

China has also succeeded in greatly expanding guidance in the use of information. China has a policy that all children must go to school for a minimum of nine years, and talented individuals are given encouragement and financial support to get further education. The literacy rate is about 90 percent or more for both sexes. Advanced education suffered during the Cultural Revolution, when dissenting intellectuals were purged and many educational institutions shut down. The post-Mao government has strongly supported growth of higher education, particularly in the sciences. In the last decade, educational institutions and enrollment in higher education have expanded by about fourfold.

The attitude of the society has changed markedly from the old days when philosophical knowledge was considered to be of the greatest value. Modern Chinese people value scientific and technological knowledge highly and are cognizant of the connection between innovation and economic prosperity. A patent office and a body of patent law were set up by 1984 and subsequently updated to meet international standards when China joined the World Trade Organization. Inventors and scientists are highly regarded by the press and the populace.

Thus in the areas of access, guidance, and rewards, China has almost caught up with Western societies. In the area of multiplicity of opportunity, the record is mixed. Many of the decentralization policies coupled with the sheer size of the society have resulted in a marked increase in the number of opportunities for education, employment, and innovation. Chinese students attend universities around the world, and Chinese businessmen operate worldwide, so both groups have access to the latest technical developments. It's not clear whether potential inventors have the same opportunities at home as they would have in Western societies, but the ruling party would probably claim that as its intention. As for research institutions, a single Chinese Academy of Sciences works with technology-based companies.[4] Its members are appointed and intended to represent the best scientific minds. It has thirteen regional branches around the country.[5]

The recent government has wisely opened up much of its economy to competition, allowing successful entrepreneurs and financial backers to keep most of the profits, rather than using the profits strictly as sources of government revenue. In a society that has historically valued stability and viewed decentralization of power as a dangerous threat leading to chaos, control is still paramount. Thus, there is considerable resistance within the party to any decentralization, because that would inevitably lead to new power centers outside the party structure. How this works out in the future is anyone's guess. The decentralization could continue, and the dominant society could become more comfortable with multiple power centers, in which case a strong middle class could develop, and median income could rise to match the West. The opposite could also happen, wherein the newly powerful businessmen could be co-opted and become part of a new plutocracy, similar to what happened in the Italian city-states of the Middle Ages; if this happened, the middle class would be small, and median income would level off at some lower value. While the multiplicity of opportunity for innovators is much better than before, much needs to be done to close the gap with highly innovative societies.

When it comes to the absence of fear condition, the outlook is mixed. Practical inventions that help a company be successful are encouraged, people are encouraged to file patents, there are no religion-based constraints, and the anti-intellectual purges of the Cultural Revolution have been officially repudiated. However, memories of the purges—reinforced by the frequent posters of Mao, periodic punishments of dissidents, and the sometimes-brutal actions of powerful people in the party—stir up wariness in peoples' minds. Could the old regime reappear? Where are the boundaries of free speech on what you can and can't post online? Could today's innovative ideas be used as damning evidence sometime in the future? Could the state take away money, property, or professional positions under some circumstances not anticipated today? Considering that so many students who study abroad choose not to return, that a disturbing number of successful businessmen have second homes abroad, and that quite a few people continue to move out of China when they can afford to do so, all indicate that the people don't completely trust the government to protect the fruits of their efforts in the future. This is exacerbated by the fact that all the land is owned by the state. While such a political atmosphere doesn't prevent a modest level of innovation, it does limit the commitment and motivation of individuals. There is still a lack of transparency in the society, and control is still paramount in the minds of the government. The known high level of corruption and occasional brutal actions of party higher-ups are reminders of the possible dangers of sticking one's neck out too far. In spite of its stated commitment to ideals of free speech and a harmonious society, the party has not yet created the kind of fear-free atmosphere essential to an innovative society.

The Chinese government is attempting to provide wider access to certain types of information and training, as well as to reward business initiatives. It is not clear that it recognizes the need for a multiplicity of opportunities. It is clear that it believes it possible to maintain a docile population through intimidation and still create a prosperous society. Most Chinese seem to be willing to accept a much higher level of intimidation than Americans and Europeans. The result of this remains to be seen, but certainly it will limit the areas of knowledge where innovations will be encouraged.

Other Democracies—Independent Former Colonies

Canada, Australia, and New Zealand have histories that resemble that of the United States; they had no aristocratic history, and they were able to avoid the establishment of aristocracies and monarchic institutions. Of course, like the United States, they share a shameful history in their treatment of the indigenous people who lived in those lands before the Europeans arrived. Also, as in the United States, the indigenous people lived in areas with low population densities that prevented the imposition of an aristocratic class structure. These countries achieved independence from Great Britain a couple of centuries ago and established elective governments and democratic institutions, so it is no surprise that they have innovative populations similar to those in western Europe and the United States.

Postwar Asian Democracies

Soon after World War II, most Asian countries moved toward democratic governments; Japan, South Korea, Taiwan, India, and Singapore are among the most prominent. Japan had democracy imposed on it after the war; South Korea and Taiwan were dictatorships. All except India had strong establishment plutocracies (for example, the Korean *chaebol* and Japanese *zaibatsu*) that promoted industrialization and embraced Western science and technology, recognizing their universality and their societal and economic benefits. They evolved into prosperous democracies with legislative branches elected by the populace. India became independent in 1947 and was immediately established as a democracy. From independence to 2000, its GDP per capita rose from a subsistence level of GK$600 to over GK$2,000.

Japan has a fascinating history that defies easy description. While it has had elements of autocracy, plutocracy, dictatorship, and democracy, until the last few decades of democratic government, it was none of these. It has had emperors and shoguns who didn't have the absolute power of autocrats. Feudal lords governed the land and warred with each other, and they didn't join forces to form a single plutocracy. For much of its history, Japan had a rigid class structure. In spite of these features, Japanese craftsmen were able to quickly duplicate and adapt foreign inventions such as guns—Japanese guns were among the most advanced in the world. Japan resisted contact with Western cultures for several hundred years, but within a few years of renewed contact in the 1860s, the country adopted a number of Western institutions and embarked on its own industrial revolution. By the end of the century, it had succeeded dramatically and had become an international power. The GDP per capita grew from about GK$700 in 1870 to GK$2,800 at the beginning of World War II. After the war, democratic institutions were imposed on the country, a very unusual circumstance. These institutions have been retained since political autonomy was restored. Today Japan has one of the largest economies in the world, and its citizens are among the most affluent.

When it comes to innovation, the Japanese experience is particularly significant because it has defied what was conventional wisdom a few years ago in two respects. First, Japan was considered a copy-cat culture that was good at duplicating Western manufacturing techniques, but not at developing new ideas. It is now apparent that in the process of studying Western techniques, the Japanese developed ideas on how to improve them, which is exactly how the innovative process works. While transistors were invented in the United States, it was Japan that used them to dominate the market for television, radio, and recording devices. Japanese automakers developed methods of improving quality while keeping prices competitive, and became highly successful. Second, some Western observers considered the country's culture to be too group oriented, and not individualistic enough, to become highly innovative. Nevertheless, the Japanese found ways of rewarding innovative people, and Japan has many top-notch engineers and scientists. As one indicator, for several years its inventors have been annually awarded more American patents than any other non-US country.[6] This shows that innovative people will respond when the five essential conditions are present.

Latin American Countries

Latin American countries, from Mexico in the north to Argentina in the south, are just emerging from several centuries of brutal colonization by the Spanish and Portuguese. This legacy left most of their land and wealth in the hands of a few families and set the stage for the later military dictatorships. This inequity still severely hampers the progress of the last four decades. These countries have been held back by aristocracies that controlled the wealth and the governments and traded power among themselves. In the last few decades, the elites have been affected by the widespread recognition of the benefits of democracy, so they have bought into it, to some extent. Pushed by popular uprisings and resistance, Latin American countries have developed more democratic institutions and have established elections of public officials. The conflict between the despotic and the democratic forces continues to play out; bribes and corruption are still prevalent, but they are no longer accepted as necessary evils, but rather as actions that should be punished.

The effects of increasingly independent media, foreign tourists, travel by citizens to foreign countries, television and internet access to the experiences of people in other countries, and better education are causing changes of attitude across the population that hopefully will result in increasing democratization. People are still more likely than in the United States and western Europe to admire a dictator like Juan Peron or Pinochet and more likely to be concerned with the power of politicians than with democratic processes. Hopefully this will change as the citizens see that progress can occur without some strong leader making it happen.

Innovation was minimal under the dictatorships, so now that the five essential conditions have improved, we are seeing increased innovative activity. As the people adjust to the improving environment, we can be confident that it will continue to grow. There will probably be setbacks, with new dictators emerging, but the widespread availability of the democratizing innovations and the difficulty facing despots who would like to control the information sources provide a basis for optimism that their days in power may soon come to an end.

Other Countries

There are many other countries that have not been discussed in this book. Each country has its own story and its individual reasons why it has not yet reached the level of democracy of the Western industrialized nations or that of the benign plutocracy of some Asian nations. Some, such as many African nations, are torn by internal and tribal conflicts, some are under the thumbs of dictators, and some have not yet emerged from their colonial past. Others, however, have overcome these problems and are well on their way to becoming true democracies, examples being Turkey and many central and eastern European countries.

Most Middle Eastern countries have governments that are best described as benign plutocracies. Saudi Arabia is one example; while it has a king, its plutocracy is the large Saud family, which controls the politics and the economy of the country. It is benign in the sense that it spends a considerable amount of its oil revenues on providing social services and safety nets to its citizens. It also supports education and values science and engineering. These features are also true of a number of other Middle Eastern countries. Iran is another example of a benign plutocracy: while it has elections, the candidates are not picked by freely operating political parties. The plutocracy in this case is the Islamic clergy, who, while they don't drain the treasury to maintain opulent lifestyles, still believe they need to control the politics of the country, and this spills over into the economy and into attitudes toward innovation. The examples of Turkey and Indonesia as democracies show that the Islamic religion does not preclude democratic institutions from being established. The recent "Arab spring" developments demonstrate that the popularity of democratic institutions is virtually universal. They reduce fear, promote the accessibility of information and education, and encourage an atmosphere friendly to innovation. Hopefully they will continue to be developed in these countries.

Silicon Valley

The experience of California's Silicon Valley, where I live, is an example of how geography, intercultural contact, and the five essential conditions collectively provide a favorable environment for innovative activity. The San Francisco Bay area's location on the Pacific Ocean makes it readily accessible to Asia. The city of San Francisco is bordered by mountains and water to the west, north, and east, but it is open to the south to San Jose; this area, officially called the Santa Clara Valley, is bounded by two mountain ranges. The weather enhances the quality of life: it virtually never rains between April and November, and the temperature is rarely very cold or very hot. Nearby are beautiful hills, mountains, and beaches that encourage outdoor living. By the 1950s this area contained a string of towns in close proximity, with Stanford University centrally situated.

From the days of the gold rush, the notion of California as a land of opportunity has captured people's imaginations, so it has attracted ambitious and entrepreneurial people from around the United States, indeed from around the world. Three early enterprising individuals stand out in the local public lore: William Hewlett and David Packard, who formed the Hewlett-Packard company just prior to World War II, and William Shockley, inventor of the transistor. The accomplishments of these men and their many talented and ambitious associates showed that money and fame could be made in technology, especially electronic technology. Others soon followed, along with engineers of every specialty imaginable, accountants, salespeople, technicians, production workers, marketers, lawyers, financiers, and a wide variety of consultants. Students, inventors, and entrepreneurs flocked to the valley from China, India, Vietnam, Japan, Europe and Iran—really, from all over the world. Students from Stanford and other universities nearby found jobs, learned from them, and branched out to form new companies. Santa Clara Valley became Silicon Valley. Today it is the home of Adobe, Apple, Cisco, eBay, Google, Intel, Oracle, Sun, and Yahoo!, to name a few outstanding companies.

An entrepreneur wanting to set up a new technology company here has expert consultants in every conceivable technical and managerial area to draw on. Banks and venture capitalists, automated processor board and silicon chip assembly lines, job fairs, trade shows, not to mention meeting and eating places—everything one would need—are all within a half-hour drive. The person most familiar to me who personifies this entrepreneurial spirit is Charles Trimble, who founded Trimble

Navigation, where I worked for many years. Charles was working for Hewlett-Packard when he learned that the company was reluctant to pursue some navigation-related technology it had developed, so he bought the rights to a design for a GPS receiver. Soon he and some similarly entrepreneurial colleagues developed a receiver for civil use. It was risky because there would not be enough GPS satellites to provide continuous service for several years. He and his staff took full advantage of the opportunities in Silicon Valley and persevered through several crises, and today Trimble is a $2 billion per year company supplying a wide variety of high-quality GPS receivers and systems to civil and military customers.

From the perspective of the five essential conditions for innovation, Silicon Valley excels. It is one of the most fear-free places in the world for innovative minds. Foreign students, businessmen, and technical professionals are welcome to try out their talents; there is probably less fear of failure here than anywhere. With so much concentration of brainpower here, information, education, mentoring, and consultation are immediately accessible. Innovative people are rewarded with jobs, patents, and stock options; they have opportunities to start their own businesses if they wish. The opportunities are abundant; a talented person can find work at any number of companies and in any number of technology areas. So Silicon Valley is a virtual showcase for demonstrating the power of the five essential conditions. It is a mini-society that provides them all in abundance.

The Significance of Societal Size

In chapter 7 it was noted that potential market size has an impact on inventive activity. Although trade barriers are gradually diminishing, the potential market for many inventions is initially limited to the country of origin. If a new product proves successful there, typically the inventor, company, or financial supporters then look to adapt it to foreign users with different languages, electrical plugs, national standards, or other regional differences. Different inventions require different amounts of upfront investment, so the potential market has to be large enough to cover initial costs and make a good profit. This means that inventors and entrepreneurs in larger countries have two distinct advantages: there is often a larger initial market, which attracts innovators, and it is possible to be more ambitious in designing new products, because the larger market makes it possible to spend more money during development. This has given the United States—with its large population, common language, and common standards—an advantage in the past; even the commonalities in the culture worked to its advantage. A new product developed for US consumer use has immediate access to a large market. In a smaller country, the potential markets for many inventions are initially smaller, so it somewhat reduces the motivation for innovative activity.

This picture is changing. More people are learning English and becoming multilingual, the internet is dramatically opening up markets for some types of innovative activity, worldwide standards are becoming increasingly common, and products are increasingly being developed in one country and manufactured in another. Also, products are being developed using international teams, so that they are initially designed for a wider market. This means that Americans can expect more competition from inventions originating in other countries.

Societal size also means that the number of opportunities is fewer for innovative people in smaller countries, simply because there are fewer industries. This, too, is changing, because regional alliances like the European Union are making possible increasing mobility: people can study and work in foreign countries, so they have more opportunities than before. With modern communications and the internet, it is easier to establish and maintain contact with people and companies in foreign countries. So while societal size continues to matter, it is becoming less important with globalization.

The Decreasing Importance of Natural Resources

I was struck by the wisdom of a lecture by Shimon Peres[7] on the significant ideas of the twentieth century. He stated that a very significant development in the first half of the century was the end of imperialism as a viable concept. During that time the notion that one country can benefit itself by controlling the resources and the people of another country became discredited. The costs outweighed the benefits as never before: maintaining a large occupying army was just too expensive. It became generally recognized that economic trade between the same countries as equals can get the same benefits without the human and financial costs. Within a few years of the end of World War II, the empires of Great Britain, France, Holland, and Belgium ceased to exist. The Soviet empire came to an end before the end of the century. American hegemony in the Western Hemisphere was greatly reduced. Today there are just vestiges of imperialism, although scars of old still remain.

Peres also said that by the second half of the century, it had become evident that the economic success of a society was no longer as dependent on the natural resources within its boundaries as it had been historically. He cited the economic success of Japan and Israel, neither of which has significant natural resources; to these, one could add South Korea and Taiwan. Each country imports raw materials and balances (or exceeds) them with exports of finished goods. On the other side of the coin is Russia, blessed with an abundance of natural resources, but struggling to find a way to build a viable economy. Africa has more than half of the world's arable land, but it currently produces only a small fraction of the world's agricultural output. This is not to say that resources no longer matter, but rather that it is possible for a country to thrive without them. Governments of countries that are highly dependent on a single resource like oil are in a precarious position in the long term. Recognizing this, some governments (e.g., Norway and the United Arab Emirates) are making preparations for the day the oil runs out.

The Ascendance of the Democratic Concept

We have seen that a typical evolution of established societies was from autocratic, monarchic, or dictatorial to plutocratic, then transitioning to democratic. During their plutocratic periods, the societies went through industrialization and economic expansion that brought great prosperity to at least some portion of the population. While democratization was strongly resisted by many of the wealthy beneficiaries of this process, there were enough enlightened people in power and there was enough pressure from the populace that democratic institutions such as legislative bodies, property laws, and court systems were established. Usually these institutions were initially set up to benefit certain classes or economic interests, but over time they were expanded to include more and more of the general populace. As more of the population became less fearful and more aware of their rights and opportunities, they insisted on having increasing control of their own lives. Elements of the population fighting for specific rights originated thousands of protracted struggles. They were successful because even the people in power had to admit to the justice of their causes. There are few examples in history of rights being handed down unsolicited from the powerful to the populace.

The democratic concept involves more than just electing some officials. It also involves governmental and societal checks and balances to prevent corruption and excessive concentration of power. It involves supporting different political parties that have a realistic chance of gaining power and can function without fear. It involves an independent free press that can criticize the government and its actions. It includes the ideas of transparency and accountability in transactions between citizens and institutions, both governmental and private. It involves, in the words of the first amendment to the US Constitution, the "right of people peaceably to assemble, and to petition the Government for a redress of grievances." A familiar slogan that describes a false democracy is "one man, one vote, one time," (i.e., an election in which a dictator gains power, and then gets rid of the electoral process); this has happened in the past and shows that, without the associated democratic institutions, the vote by itself is very precarious.[8] There have been numerous cases in plutocracies and dictatorships where strong leaders have dismissed democratically elected parliaments because the checks on the executive branch of government weren't strong enough.

The democratizing innovations discussed here have been crucial to the ascendance of the democratic concept. Newspapers and news magazines made possible by the printing press were crucial in getting opposing viewpoints out to the public. Radio and television expanded this capability. Modern electronic technologies have provided added techniques: cameras are small and easy to conceal, and many cell phones have them. Between broadcast television news and the internet, it is becoming increasingly difficult for a despot to keep secret the actions of the police and military. Television broadcasts have enabled most people of the world to see the difference in living conditions between the citizens in democratic countries and those in nondemocratic ones, a strong factor in the downfall of the Soviet Union. Bland assurances of rulers and politicians are more than ever subject to dispute and refutation. Person-to-person communications provide information that belies the claims of government-controlled media. While benign plutocracies have been successful in raising the standard of living of their citizens with varying degrees, most people living in these countries would like to gain more individual rights and democratic institutions. They see that although problems with corruption and inequality still exist in democratic countries, these problems are much worse elsewhere.

Interestingly, democratizing innovations appear to have affected the ruling classes as well. Historically, they went to great lengths to shield themselves from contact with the lower classes, walling themselves up in great estates and mansions. Consider the story of Siddhartha Gautama: he was born into royalty, and it is said that his father tried to prevent him from seeing the suffering of the common people. It was when he came upon a sick old man that he began to think differently about society and eventually founded a new religion, Buddhism. In the cities and towns of Europe, it became increasingly difficult for the rich to avoid contact with the poor. While the monarchs continued to live in their palaces and the landed aristocracy in their country mansions, the merchants, businessmen, and town officials had to deal with the poor on a daily basis. The spirit of nationalism made it more difficult to ignore certain parts of society and to consider them irrelevant. Once radio and television appeared on the scene, living conditions in all strata of society were on display to everyone. More recently, with small, concealable cameras and cell phones, pictures and videos get broadcast over television and the internet, and everyone sees them. Rich and powerful people are human beings as capable of compassion as anyone else, so they respond to the images they see. As just one of countless examples, the American civil rights movement succeeded once the vivid television images of injustices and racial hatred were seen by the general populace and the establishment.

Today it is clear that democracies provide and protect individual rights better than any other form of government. The existence of democratic institutions, including a free press, is considered by most people everywhere to be the gold standard for a free people. Plutocracies recognize this and use words and phrases like "freedom," "democracy," and "the will of the people" while attempting to convince their people that for cultural reasons they must pursue a different course than the democracies of the West. This is a ruse, and the people who live in these societies increasingly recognize it; the real issue is control. Control is never easily relinquished, and there are all manner of excuses for retaining it: national security, religion, culture, and the threat of chaos are the most common. While this need to retain control affects all five of the essential conditions for innovation, its primary effects are to maintain a level of fear in the society and to justify limiting the multiplicity of opportunities to the citizenry.

The main problem with all plutocracies, benign or predatory, establishment or oligarchic, is that the perceived need for control by the plutocrats is at odds with the essential conditions for innovation. Democracies, more than other political structures, have succeeded in establishing institutions that greatly dilute this control.

18. Applying What We Have Learned

Promoting the five essential conditions for innovation; encouraging young innovators; avoiding complacency and cultural biases; have we reached the limits of democratizing innovations?

The book has emphasized how governments and political climate have impacted innovative activity by encouraging or discouraging the five essential conditions. It is clear that these same conditions can be beneficially applied in our institutions, companies, schools, and families. The following sections provide some suggestions for doing this, based on my own experiences. The reader, having lived through her or his own unique life experiences, will no doubt be able to formulate many other suggestions for incorporating these factors into our lives and our societies. Later sections will address some pitfalls that we as a people need to avoid if we are to build and maintain vibrant societies that encourage innovation. The last section will address the question of whether, in the current age of the ubiquitous and instantaneous internet and the boundless popular enthusiasm for innovative activity, we have reached some limits in what technology can do to support democratization.

What Governments Can Do

Democracies have succeeded in reducing governmentally imposed fear that would inhibit technological or scientific innovation. Benign plutocracies like China and Saudi Arabia have attempted to do the same while maintaining control of political thought and discourse, but they have achieved only limited success. The problem is that political control implies economic control, and innovations that would make significant improvements in the economy will necessarily threaten some plutocrats' interests. The people living in these societies understand this, and they are afraid to push too hard, instead contenting themselves with acting incrementally within the system. The situation is even worse in dictatorships and predatory plutocracies. Our hope is that eventually the democratizing forces in society will prevail. Certainly the last fifty years have seen the emergence of new democracies, and the establishment of democratic institutions within other forms of government, in response to the almost universal recognition of the benefits of democracy.

We live at a time when the techniques and tools are in place to provide everyone immediate access to an unbelievably large amount of information. Modern internet websites provide ready access to useful information; one example is Wikipedia,[1] an incredible online tool that I frequently consulted to check historical events, people, and places while writing this book. Those of us lucky enough to be able to utilize these facilities can explore in hours ideas that as recently as thirty years ago would have taken weeks. Libraries and professional organizations have mountains of books and articles containing experimental data, interpretations, and historical events. The problem is that not everyone has equal access to this information; people in many poor countries and rural areas don't yet have adequate access to good libraries, books, or the internet. Some governments place limits on available information, under the guise of avoiding chaos. So while the tools exist to provide universal access to information, this is still a work in process. Fortunately, the benefits are obvious enough to everyone that access is increasing around the world. Governments can help by expanding internet, television, radio, and cellular telephone infrastructures; they can also do more to get the corresponding consumer equipment in the hands of the public.

Democracies have quite a head start in establishing universities, research centers, and public education. This has meant that creative and innovative people in other countries frequently leave their homes to attend these institutions and usually choose to stay in their adopted countries, where they find work using their talents. Other governments wanting to establish such institutions face three difficulties: (1) democratic countries already have successful institutions, which attract many of the best students, (2) work is usually available for these students in the democratic countries but not in their own, and (3) attracting the best researchers and instructors to staff the schools and research centers in nondemocratic countries is difficult. In plutocracies the further problem is that preferential treatment is usually given to the sons and daughters of the upper classes, which drags down quality and discourages the populace. If these governments want to encourage innovative activity in their societies, educational opportunities should be provided that realistically are accessible to everyone, men and women alike; and equally important, employment opportunities must follow the education.

Governments can reward scientific innovation by establishing research programs at universities, and many have been at least modestly successful in doing so. This effort has been enhanced by the widespread access to knowledge through the internet and contact between scientists that we enjoy now. This has enabled researchers around the world to specialize in topics of more local relevance, but which are of interest to the general scientific community, such as archeological explorations. Technological innovation intended for an immediate economic impact is a different story, because to make significant headway requires a considerable infrastructure that provides patents and business opportunities. This can't be built overnight. Countries that have achieved some success in developing this infrastructure, like Japan, India, and China, have allowed many of their promising people to participate in companies in the United States and Europe to learn skills and techniques, and then have provided them with opportunities back home to utilize their new knowledge. The risk for them is that innovative people will want to remain in their new countries. This happened for a long time, but as opportunities have become available in their home countries in recent years, many innovative people have returned.

Much technological innovation today is driven by product development opportunities, the dream being to develop a "killer application" that reaches a huge consumer market. Not everyone is so motivated, however, and we have seen that most innovators are motivated more by the joy of innovation and seeing their efforts put into practice than by getting rich—entrepreneurs and businessmen are the ones motivated by profits. Many potential innovators would like to improve conditions for their countrymen, while companies understandably want to focus on product improvements that maximize profits. Thus, pharmaceutical companies are more motivated to spend money developing drugs that address erectile dysfunction, which are highly profitable, than those that address AIDS or diseases that affect people in poor or rural areas, which are not so profitable. In fact, there is a huge need to develop products and services for heretofore-underserved people. One challenge is to provide low-cost versions of existing products and services, described in the book *Jugaad Innovation* (2012). The authors make a compelling case that much can be done by individual entrepreneurs, small businesses, and even large corporations. One example they describe[2] that illustrates this is a small, portable infant warmer developed in Stanford University's d.School course, Entrepreneurial Design for Extreme Affordability.[3] It was designed to keep premature babies warm for several hours, long enough to enable parents to carry them to hospitals from their villages or farms, which often don't have electricity. The warmers cost a fraction of the price of a standard incubator. Another example is microfinancing, such as that developed by Muhammad Yunus (1999), where inexpensive short-term loans are provided to people in poor rural areas to enable them to purchase materials for their handmade products without paying exorbitant amounts to middlemen. Governments can encourage such efforts by providing financial support and other rewards and removing regulations written for large institutions that could discourage small-scale efforts.

The most difficult thing for plutocratic governments is to provide a truly competitive environment for business, which is crucial for inventive activity. It requires offices that grant patents on the basis of merit, laws that punish corruption and bribery and that discourage political favoritism, a press free from political interference that can publicize wrongdoing, government offices that award merit-based contracts, and independent courts that enforce laws fairly. Few plutocratic governments have these, which is one of the main reasons democracies outperform them. While such institutions can't realistically be put in place overnight, the goal of any government that purports to increase innovative activity should be to move in this direction, sooner rather than later. These issues can be seen playing out in China today, where corruption is recognized as a problem at the highest levels of government, and most protests in the provinces involve local corruption as an issue.

What Organizations Can Do

Organizations such as businesses, governmental agencies, universities, and charitable foundations have to continually balance the issues of control and innovative activity. In many aspects of its activity, each organization has to have a consistent set of policies, speak with one voice, and make decisions that everyone has to live with. To be effective, an organization should also have a variety of avenues for addressing new situations that inevitably occur. While many people are quite content to have their jobs prescribed and to be told what to do, most people want to have some influence on organizational and strategic decisions. Those who want a lot of influence gravitate toward managerial positions; their styles range from control freaks to consensus managers, with all shades in between. Control freaks are often effective in organizations that don't need much in the way of innovative activity, while more engaging managers do better in organizations that do need it. A control freak in a high-innovation business would likely lose all the good employees who would want their views to be considered. On the other hand, a consensus manager could drive a low-innovation business into the ground if the nature of the business required a lot of of micromanaging. That said, there is a need for innovation in virtually any organization at every level, and a good manager will find ways to tap into the innovative talents of the employees.

Applying the first four of the five essential conditions is quite obvious. Reducing fear means creating a work environment where people feel free to propose their ideas. Providing access to information is relatively easy these days. Guidance primarily comes in the form of mentoring and contact between peers and providing forums where ideas can be discussed. Some employee-training courses are also helpful and make the staff feel more involved and respected. Rewards in the form of awards, bonuses, and raises are fundamental, but respect is ultimately the most important. If employees feel that their efforts are appreciated and their views respected, they will usually stick around.

The fifth essential condition, providing a multiplicity of opportunity, is underutilized in most organizations. We know that people are motivated by competition, and companies usually motivate employees to compete externally in the marketplace. Internal competition is usually discouraged because of the potential for hard feelings and disruption of the work. Properly handled, however, it could be a means of developing product and strategy proposals that are more thoroughly analyzed. The competitive spirit is a powerful motivator, and it could be used to better advantage.

What Parents Can Do

Parents who read this book may want some ideas on how to encourage innovative thinking in their children. At this point I will give some opinions on the subject, but with the caveat that I have no training in education, just my experience as an engineer and parent. I fully expect that everyone will have their own views on the subject, based on their own experiences; the material presented here can stimulate thinking about encouraging innovative activity. I hope that teachers will find this discussion useful as well.

It is clear that absence of fear in the home is important in fostering innovation. The primary fear for a child is appearing stupid to parents, siblings, and friends. Creating an environment at home that encourages curiosity and the expression of ideas and avoids making fun of children's statements or denigrating their judgment is important; and when children are exposed to such negative statements, which happens a lot, parents can help their children deal with the situation without losing confidence in themselves.

Providing access to information today has become incredibly easy for most educated people. Information is accessible today on all sorts of topics and at speeds that even a few decades ago would have been unthinkable. The potential of the present digital-age tools has already proved to be immense; search engines have progressed to the point that it is relatively easy to research virtually any subject imaginable. This has actually resulted in "information overload": there is so much information available, but its quality is quite uneven, and frequently erroneous. A major problem today is how to filter out erroneous and irrelevant information. The internet today is extremely good at obtaining factual data, and as it evolves it will become increasingly useful for the formulation of concepts and ideas. Parents would be well-advised to become knowledgeable in using and evaluating the new tools and to keep up with new developments in this rapidly changing area. Then they can make sure the available information is relevant.

At the same time, the old-fashioned book should not be ignored. Books enable authors to more fully develop their ideas and concepts and to provide useful background material that can stimulate inventive minds. For example, compare the biographical summary that one reads in a Wikipedia article with a biographical novel: the article will provide useful facts, but the novel will bring a vividness and accessibility to that person that will excite young minds and give them the message that they too can accomplish significant things. I tried to convey this in chapter 15, which dealt with the lives of some great innovators who found ways of dealing with difficult situations so they could pursue their passions.

Guidance in the use of information should involve encouragement and confidence building; at the same time, it should be made clear that really good ideas are rare and that the first ideas to come to one's mind are usually quite naive or wrong.[4] The trick is to strike a balance between over-praising and over-rewarding a child's ideas on the one hand, which encourages sloppy thinking, and denigrating them on the other, which discourages curiosity. Young minds should be challenged in such a way that a bad idea is not seen as a failure but as a learning experience that sharpens and improves the next one.

Parents can provide a crucial role in their children's education by providing them with experiences, conversations, and television programs that expand their understanding of the world. As adults we tend to forget that our associations with words are far richer and nuanced than children's associations with those words. Take the word "grain" as an example. Children's associations may be limited to what they see on a box of cereal, possibly with a picture of a field. Most adults will picture many different kinds of grain from different parts of the world. They will picture the growth of different grains from planting to harvesting, and they will be aware of the requirements for different amounts of water, different growing seasons, and different climates. They will picture grains in a variety of breads and pastries, even in ethanol. What children learn is limited by their prior knowledge and experience, so materials prepared for young students should aim to broaden and fill out these associations before more complex material is presented. Much of this kind of knowledge is acquired outside of textbooks. The more experiences one has, the more vivid are the images brought up by the words on the page. These images help trigger innovative ideas. Young students first learning a new subject can hardly be expected to be excited about something they barely understand, especially when the images evoked by the descriptions are sketchy and incomplete.

As children get older an important practice that needs to be encouraged is critical thinking. Critical thinking does not mean disbelieving or finding fault with what is heard or written. It means analyzing words and concepts in detail and considering context. How unambiguous and clear are the words and concepts? Are the statements opinions or facts, or some combination? If the statements claim to be factual, what is the evidence or expertise cited or referenced? Is the reasoning valid, or fallacious or sloppy? Thomas Edison said that at a very young age he took to challenging every sentence he read—he was the ultimate critical reader.[5]

Young people should be guided to not fully believe any assertion based solely on authority. Expert opinion should be sought out and studied, but with the understanding that even the experts have human limitations, and throughout history they have made confident assertions that we now know to be wrong. One of the most common human mistakes has been to say, "If the great so-and-so says this and that, who am I to disagree?" Such excessive veneration has had a debilitating effect on innovative thought.

Parents can give their children a multiplicity of opportunities by providing a wide variety of experiences, as described above. Also, children have different learning styles—some learn better through oral communication than through reading; poor readers are often quick learners if taught properly. For such students, the home can complement the classroom. Different children have different talents and skills, and they should be encouraged to develop the ones they have because success in one area stimulates success in others. Innovative children need some unstructured time to develop their creative and innovative skills. Parents who occupy all their children's spare time with activities may be subtly discouraging truly innovative activity. As with many aspects of life, there needs to be a balance, in this case between acquiring knowledge and applying it.

One of the striking things about innovative people is that they recognize when they must master new skills and obtain new information, and they go about doing so in a focused and tireless fashion. Once they have achieved some proficiency with language and mathematics, their interests often dictate what vocabulary they need and what topics and skills they want to pursue. In the process they improve their reading and math skills. Ideas are what turn them on; ideas motivate them to master the tools of learning, which otherwise are just sterile and pointless exercises.

People have different propensities for innovation from an early age, perhaps from birth. However, there are techniques that can tap into, stimulate, and develop the potentials that children do have. The development of such techniques is an active area of educational research today. Parents can provide an environment with ready access to information, foster curiosity and critical thinking, and avoid discouraging potential innovators. Such an environment encourages all sorts of creative activity and, most of all, brings joy and fulfillment into everyone's lives.

The Fallacy of the Best

There is a widespread fallacy that exists even today—namely, the concept of "the best." Governments seek to set up the best institutions, the best universities, and the best research centers. Universities want to establish the best departments. People want to listen to the best expert on a given subject. Companies want to hire the best corporate lawyers, CEOs, and other officers. This is a natural desire, and the same urge has led us to look to the ancients, but it has a trap. We have seen how respect for brilliant historical figures grew to veneration and even to worship, and how people forget that even the most talented and knowledgeable geniuses have flaws and blind spots. Such historical figures were limited by their knowledge at the time and by the sources of information they could access. In addition, they were accorded so much praise that many of them began to believe in their own infallibility and became careless. We should strive to fully understand the concepts of the great thinkers, but once we do, we should challenge everything they said and honor the motto of the Royal Society, "on no one's word." Societies should not attempt to set up a single-best institution or a single-best university in spite of how reasonable it might seem. The institutions will always have departments that are headed by a "best" researcher or scholar, and this person will always have biases and limitations, no matter how well qualified and esteemed. This same expert will inevitably resist the introduction of ideas different from his or her own, which will subtly discourage white-collar innovation. By contrast, a multiplicity of universities will result in a competition of ideas, a blossoming of new subjects, and a healthy fear of being left out of new developments. A prime example is the United States, which was fortunate to have a large land area and population, enough that it eventually established many centers of excellence. Today, an aspiring engineering student can apply to the University of Michigan (my alma mater), MIT, Stanford University, Princeton University, the University of Illinois, or Caltech, to name a few top technical schools. Smaller countries don't have the luxury of so many choices, but even they would be better off to establish several smaller universities than one "best" one. It is noteworthy that the European Union has recently made it easier for students to attend universities in any of the member countries; this will encourage innovation and creativity.

The fallacy of "the best" also manifests itself in the false notion that a highly centralized system could work if it had access to the best tools and information; then economies could be centrally controlled and innovation given direction. John Kay refutes this very well: "Because most decisions are wrong and most experiments fail, it is also tempting to believe that we could manage businesses and states much better if we only assembled sufficient information, cleverer people, and debated the issues at length."[6] Innovation thrives best in an environment that is not centrally directed or controlled.

Over-centralization in a society or its institutions leads to stagnation and the discouragement of innovation. Attempts to create a single best institution or organization will actually limit innovative activity; life is too complex and rich to be tied down to a single best of anything worthwhile.

Cultural Superiority

Cultural pride can inhibit innovation if taken to an extreme, and this has frequently occurred in history. All of us want to believe that our society is the best, and we always have. The military successes of powerful empires made them highly susceptible to this attitude; our ruler, our armies, our religion, or our culture is the best, and we reflect this pride onto ourselves. In retrospect, declarations of this attitude often seem silly and unjustified (think of Shelley's poem of the broken statue of Ozymandias[7] in the desert), but it is just as prevalent today and holds the same dangers. This is because it leads to complacency and an attitude of "we're successful because we're superior." This belief in the superiority of one's own society or culture usually leads to underestimating the capabilities of others. It was commonplace among the upper classes in Spain and Portugal in the sixteenth and seventeenth centuries to brag about their superiority.[8] With the wealth they had plundered from the Americas, their upper classes were able to purchase any goods they wanted from anywhere in the world, and their industries, which had been strong, became increasingly depleted. Much of their wealth was soon dissipated, and by the mid-nineteenth century, Spain and Portugal were relatively poor countries. From the perspective of people living at the time, this fall from grace was quite gradual, and there was always hope that it would be reversed—after all, these were superior, enlightened societies, run by enlightened noblemen. We all have difficulty seeing our own culture as very limited spatially and temporally; we tend to see other places and times through our own small window of experience. In reality, however, the rest of the world is a very different place from ours, and in the past people lived in very different environments.

One might think that the historical record would be relatively clear on who did what, but it was illuminating to me to discover how much disagreement there is between various authors. Writers seem to have difficulty being objective, (i.e., transcending their natural feelings of nationalistic and cultural superiority). As just one example among thousands, consider the compass, with counterclaims by Europeans and Chinese over who invented it. It was clearly documented first in China, and the record is pretty clear that needles floating on water that pointed north/south were used there, certainly on land, and probably on ships. It is highly likely that word of this was carried from there to Italy, where someone figured out how to build one that would still work when placed on a rolling, heaving ship. It certainly revolutionized navigation, and it enabled Venetian merchant seamen to travel year-round. So was the compass invented in China or in Italy? It seems clear that the most objective answer is both, because there was considerable innovative activity in both places. In fact, the history of virtually every invention begins with an initial version having limited capability, followed by many changes: inventing improvements, extending the concept to other applications, and making adaptations better suited to local conditions.

The final nail on the culture coffin was pounded in place in the last few decades, which demonstrated conclusively that within a few short years people are able to function very differently when they move to a different place. The outstanding success of Chinese, Indian, and other Asian immigrants to the United States, Canada, and Europe in business and research shows that any past lack of innovation was not due to some cultural defect, but rather to the lack of opportunities in their home countries, where the five essential conditions for innovation were not as strongly supported. When people from one culture move into a supportive environment, they can rapidly become more innovative.

Have We Reached Limits in the Development of Democratizing Innovations?

An interesting recent occurrence is that some features of democratizing innovations have reached their useful limits. For example, we have reached the limits of useful speeds of transportation. In a few hours, aircraft can take us anywhere in the world, and the experience of the supersonic Concord demonstrated that there is limited economic utility to greater speeds. Likewise with personal automobiles: while they are capable of traveling faster the road infrastructure places limits on feasible speeds. Flying cars and personal helicopters are still just the pipedreams of sixty years ago. Some high-speed trains have proven viable, but only in certain corridors. We have reached the point that while faster means of transportation are technically possible, there is little economic benefit of saving a few minutes on land or a few hours by air. The same is true of ocean travel.

In the distribution of information, we have reached the point where data can be transmitted to users in a few seconds; except when large amounts of data are required, there is little benefit to making the speed of transmission faster. The quantity of information that can be transmitted can and will increase significantly in the future, but even there the usefulness is limited by the ability of people to absorb and utilize the information. While one can download a long novel or textbook to a computer or tablet in a few seconds, it still takes hours to read and comprehend.

In communications, it is now possible to talk to anyone around the world and exchange data very quickly. Innovations will continue to increase the volume of data, but faster speeds will be of little benefit other than to accommodate this increase.

In entertainment, the point has now been reached that people can download the latest song or movie very quickly or review exciting athletic performances. They can be played with high fidelity on a large screen or a portable device. Recordings can easily be made of every kind of entertainment for later use. Theaters display on large screens with surround sound live events, like operas, that provide experiences close to live performances; in some ways, they are even better, since in live performances the audience is more distant from the performers. People will soon be able to record their personal events with high-quality displays and sound.

There are still significant improvements to be made, in particular to make available universally what is now in use only by a fortunate minority in many areas: transportation, distribution of information, communication, and entertainment. These improvements will require an expanded infrastructure and cheaper user rates. Transportation can be made safer, more reliable, and more convenient and will surely include self-driving vehicles. The potential uses that can be made of our ability to quickly create and move huge amounts of data have just begun to be explored. In medicine, in particular, the potential of currently identifiable technologies to sense significant body parameters, characterize them, transmit them to diagnostic centers, diagnose conditions, and devise and perform corrective procedures is immense. Starvation could become a thing of the past if we have the will to make it happen: food supplies and susceptible areas could be forecast and mapped and a distribution network quickly created to get appropriate food from the sources to the people in need. Epidemics and diseases could be similarly treated. The field of economics could become more exact and predictive as more precise data are gathered and better models are created; these could potentially be used to eliminate poverty and reduce economic volatility.

While some limits have been reached in the democratizing technologies, it is also true that we are in the early stage of their utilization. We are witnessing an impressive expansion of both the number of people using them and the ways they are being used. The impact of the printing press lasted for centuries, and we can expect the same from the recent technological advances in communications. The capabilities are new, and more and more people are using them.

The discussion so far has emphasized democratizing effects, but it must be acknowledged that these same technologies can be used for oppressive purposes. Despotic governments will attempt to limit the public's use of transportation, access to information, and communications with each other. They will continue to find ways to use these technologies to spy on their citizens, and they will use the massive amounts of data available on the internet to limit, intimidate, or prohibit activity that should be widely allowed in a free society. Even democratic governments are participating in these activities, using national security as their justification, a power that can be easily abused. Furthermore, the threats can come from sources other than governments; the accumulation of massive amounts of personal data by commercial interests places huge power in the hands of a few people whose interests can be contrary to those of the public.[9] It is not possible to predict whether the democratizing effects of technology will ultimately prevail, or whether control-obsessed leaders and other powerful people will succeed in negating them.

Final Comments

As I stated in the introduction, I had found that most of the conditions cited by historians for furthering innovation were somewhat remote from my own experience with innovative people, that there was a gap between the societal-level conditions cited and how they were experienced by the individuals involved. I believe the five essential conditions I developed bridge this gap, because they apply both at the societal level and to the individual experience. The feeling of freedom to explore, to be able to get relevant information and an appropriate education, to be potentially rewarded for one's efforts, and to have options—these all make sense at every level.

Innovation has now reached a level of acceptance throughout the world that would have been almost unimaginable only a few hundred years ago, when change of any kind was resisted not only by people in power but even by average people who feared that they might lose what little they had. Innovation has proved to be the engine of prosperity. While the explosion of innovation was enabled by a fortuitous combination of political and demographic events, it has also contributed to its own growth, to prosperity, and to democratic freedoms.

In the high-tech world, advances have been breathtaking; the hardware of distributing and communicating information today has been developed far beyond the software. That is, we have just begun to tap the possibilities of how to utilize these magnificent capabilities. There are endless possibilities for formulating and applying new information, in medicine and biology in particular, but also in education, the visual arts, music, and the world of entertainment.

We can see the real possibility that someday everyone will have access to the knowledge of the world and will be given the training to use it effectively. A whole new generation is looking for innovations that will make this happen, taking capabilities that are available now to the fortunate minority and adapting them for low-infrastructure environments and making them inexpensive and widely available. They are looking for ways to provide basic necessities, medical care, and education, as well as entertainment and other ways of bringing vitality and meaning to everyone.

The dangers that have always been present are still here. There are still power-hungry plutocrats and despots who will try to control information and people. They will attempt to use the democratizing innovations to extend their control and power. New imperialists will build new weapons and try to take over other countries. Religious fanatics will try to impose their views on others and control societies. These reactionary forces have been operating throughout human history, and they are still present. The good news is that, over the last hundred years, these forces have been seriously weakened; the Hitlers, Stalins, and Maos have not been replaced with younger versions of themselves, but by more benign and very different governmental systems. Dictatorships of South America, Asia, and Africa have been replaced by plutocracies and democracies. While we continually see violence on our TV screens, mass violence has actually measurably decreased in the last hundred years. Much of this is due to the democratizing innovations that have taken place and have made people less susceptible to demagoguery. May they continue to do so!

Bibliography

Aczel, Amir D. 2001. *The Riddle of the Compass—The Invention that Changed the World.* San Diego: Harcourt, Inc. [The book sheds light on the environment of the merchants of the Italian city-states at the height of their far-flung trading network.]

Acemoglu, Daron, and James Robinson. 2012. *Why Nations Fail—The Origins of Power, Prosperity, and Poverty.* Crown Publishing Company. [This book is full of excellent historical accounts, including descriptions of Venice, South America, and early colonial America.]

Agricola, Georgius. 1556. *De Re Metallica.* Translated by H. C. and L. H. Hoover, 1950. New York: Dover Publications. [Yes, it was former US President Herbert Hoover and his wife who translated this book from the Latin. It is full of plates, showing that the ancients were well aware that "a picture is worth a thousand words." It also suggests that pictures like these were common long before the invention of the printing press.]

Bocquet, Bill. 2003. "Un Savant Oublié: Philippe de Girard (1775–1845)." *Le Journal de la Vieille France*, no. 54. Chalaines, France. [The title translates as "A Forgotten Scientist," which is not accurate because de Girard was an inventor of useful technology, not a scientist per se; however, in his inventions he showed that he had a firm grasp of the scientific knowledge at the time. My friends Claude and Dominique Berger-Ceccaldi found this reference for me.]

Bodanis, David. 2000. *$E=mc^2$—A Biography of the World's Most Famous Equation.* New York: Berkeley Publishing Group. [This delightful book brings a human perspective to the scientists who participated in the development of the concepts of energy and mass and the connection between them via the speed of light.]

-------------------. 2006. *Passionate Minds.* New York: Crown Publishers. [Another delightful book that chronicles the passions, both physical and mental, of Voltaire, who was far more than a playwright, and Emilie du Châtelet, one of the earliest female scientists.]

Burns, Edward McNall, and Philip Lee Ralph. 1958. *World Civilizations from Ancient to Contemporary, Volume 1.* W. W. Norton &

Company, Inc. [This is a standard textbook, and it has been revised several times.]

Cahill, Thomas. 2003. *Sailing the Wine-Dark Sea—Why the Greeks Matter*. New York: Nan A. Talese, Doubleday. [This is a well-written history of the Greek experience by a master historian.]

Conner, Clifford D. 2005. *A People's History of Science—Miners, Midwives, and "Low Mechanicks"*. New York: Nation Books. [The author does a great job of relating the blood, sweat, and tears of unheralded innovators, the ones I refer to as blue-collar, who generated the experimental evidence behind many of our scientific concepts and discoveries. He shows that they were intellectually far more than simple artisans who conducted "shelf runs" or random experiments.]

Diamond, Jared. 1997. *Guns, Germs, and Steel—The Fates of Human Societies*. New York: W. W. Norton. [This is one of the most important books of the last few decades. He convincingly shows the influence of geography and technology on human history. In my opinion, he has been unfairly accused by some authors of claiming that geography was the sole cause of the technological superiority of the West. My reading of his book is that he is merely claiming that geography played a strong role that had previously not been adequately appreciated, a claim I believe to be correct.]

Dyer, Frank L., and Thomas C. Martin. 1929. *Edison, His Life and Inventions*. New York: Harper Brothers. [This is the standard biography of Thomas Edison, and it contains a number of personal quotes and stories. I concentrated on the material from Edison's early life because I thought it was most relevant to the development of a great innovator.]

Fermi, Laura and Gilberto Bernardini. 1961. *Galileo and the Scientific Revolution*. New York: Basic Books, Inc. [There are many biographies of Galileo, but I thought this one would be especially good, as it was written by two Italians: an internationally recognized scientist and a writer who was married to Enrico Fermi, the Nobel Prize–winning physicist.]

Franklin, Benjamin. 1771. *The Autobiography of Benjamin Franklin*. Random House (1944). [It is difficult to find the words to express my admiration for this man. He went from humble origins to become a famous scientist and inventor and a Founding Father of his country, risking his neck in the process. In his late years, he somehow learned to navigate the hugely complex labyrinth of French political society, and he succeeded in convincing France to assist the colonies at a critical time in American history.]

Groningen Growth and Development Center. 2013. Website at http://www.rug.nl/research/ggdc/.
[The ambitious concept of providing a standard monetary reference for computing gross domestic product (GDP) across time and in the different areas of the world was developed here; it is called the Geary-Khamis dollar (GK$) after two economists who developed the concept. It is referenced to the value of the US dollar in the year 2000. It is far from a perfect indicator, because it does not directly take into account the distribution of income within the society, but it is indicative. A better measure would be median income, but that information is not available. With this caveat, GDP per capita is a credible indicator of the general prosperity of a society, and the landmark series of studies under the guidance of Angus Maddison at the Groningen Growth and Development Centre of the University of Groningen that generated the database is a treasure trove of information.]

Jacobs, Jane. 1992. *Systems of Survival—A Dialogue on the Moral Foundations of Commerce and Politics*. New York: Vintage Books. [While I reference her work only once, her concepts of two different moral syndromes resonate with the tenets of my book. One moral syndrome emphasizes loyalty over fairness, and this accurately describes the moral rules that guided autocracies in the past and dictatorships today; the second one was introduced by merchants who found that honoring contracts and hiring competent strangers were essential to conducting business successfully. These two moral syndromes are alive and well in today's societies and institutions, and the collisions between them are the stuff of much of the news of the day.]

Kaplan, Robert 2002. *The Nothing that Is: A Natural History of Zero.* New York: Oxford University Press, 82. [From today's perspective, it's hard to appreciate why the concept of zero was not well understood—after all, things go forward and backward and up and down, and we have negative and positive numbers. So it's obvious that there has to be a "zero" state—no motion—and a state neither positive nor negative. We have particular difficulty understanding why religious considerations got involved. Kaplan's book addresses this.]

Kay, John. 2004. *Culture and Prosperity—The Truth About Markets—Why Some Nations Are Rich But Most Remain Poor.* New York: HarperCollins Publishers, Inc. [This is an important book that has not achieved the recognition it deserves. It should be required reading for every economist, politician, and businessman; actually, everyone should read it. He is the only writer I have read that I would say understands economics at both the macro (big picture) and micro (detailed) levels. Written before the recent worldwide economic recession, this book lays out the dangers of "moral hazard" in a way that now seems prophetic.]

Landes, David S. 2003. *The Wealth and Poverty of Nations—Why Some Are So Rich and Others So Poor.* New York: W. W. Norton & Company. [This is an important, well-written book by a master historian on economic history and contains a host of good historical information. The author gives considerable attention to innovation through the ages and its impact on the economies of the world.]

Lewis, Bernard. 2002. *What Went Wrong? Western Impact and Middle Eastern Response.* New York: Oxford University Press. [The author does an excellent job of describing the gradual decline of the once-open and innovative Islamic society, and showing how it happened.]

Linklater, Andro. 2002. *Measuring of America—How an Untamed Wilderness Shaped the United States and Fulfilled the Promise of Democracy.* New York: Walker & Company. [This book was not directly referenced, but it is important because it shows in detail how seemingly small technological advances had significant social impacts.]

Manchester, William. 1993. *A World Lit Only By Fire—The Medieval Mind and the Renaissance—Portrait of an Age.* Boston: Little, Brown and Company. [The author describes the intellectual climate of the Middle Ages and helps us understand some of the thinking processes of the time, quite different from today.]

McLellan, James E. III and Harold Dorn. 1999. *Science and Technology in World History—An Introduction.* Baltimore: John Hopkins University Press. [As the title indicates, the authors use the modern words "science" and "technology" to describe innovations of the past rather than "white-collar" and "blue-collar." However, their excellent history does show how these types of innovations initially involved different communities, and how they eventually came together.]

Menocal, Maria Rosa. 2002. *The Ornament of the World—How Muslims, Jews, and Christians Created a Culture of Tolerance in Medieval Spain*. Little, Brown and Company. [This is a delightful book that describes what the subtitle indicates. It provides a perspective that clearly shows that the notion most of us have had of the "hated Moors" was not true for most of the people who lived there. Medieval Spain was a comparatively peaceful place where intellectuals thrived, and people from all three religions got along reasonably well.]

Menzies, Gavin. 2003. *1421—The Year China Discovered America*. New York: HarperCollins Publishers, Inc. [This book may not be taken seriously by academic historians, but it's a fun and thought-provoking read.]

Mernissi, Fatima. 1992. *Islam and Democracy—Fear of the Modern World*. Cambridge, Massachusetts: Perseus Books. [This book, written by a Moroccan sociologist, addresses the mixture of fear and admiration that many Muslims feel toward the West. The author analyzes these feelings in some detail and concludes that scientific and technological ideas are compatible with Islamic beliefs.]

Micklethwait, John and Adrian Woodridge. 2005. *The Company—A Short History of a Revolutionary Idea*. New York: The Modern Library. [We don't normally think of a type of organization as being innovative, but the idea of a company, and later its embodiment as a limited liability company, was indeed an innovation that proved to have a huge impact on product inventions and the economies of the world.]

Needham, Joseph. 1993. *Science and Civilization in China, Volume VII, Part 2*. Cambridge University Press. [I don't pretend to have read this twenty-seven-volume set, which is considered the definitive history of Chinese science and technology. However, the sheer number and quality of the Chinese innovations he describes are very impressive.]

Radjou, Navi, Jaideep Prabhu, Simone Ahuja. 2012. *Jugaad Innovation—Think Frugal, Be Flexible, Generate Breakthrough Growth*. San Francisco, CA: Jossey-Bass. [The authors make the case that while much inventive activity has focused on new products for well-to-do people, there is a whole area of potential innovative activity in developing inexpensive products for less prosperous people, one that is less dependent on the large infrastructures typical of developed countries.]

Rampal, Benjamin. 1863. *Philippe de Girard.* Paris: Jouaust et Fils. [This is the most complete biography I found of Mssr. de Girard. The librarians in Apt, Provence, found this and brought it to my attention. My account of de Girard's life in chapter 15 assumes the accuracy of this account. I recognize that from a professional historian's point of view, many of the claims may not be considered definitive, as the author provided no references. However, I believe the author was making a sincere attempt to be accurate, and the picture that emerges of this remarkable man is a consistent one.]

Rosen, William. 2010. *The Most Powerful Idea in the World: A Story of Steam, Industry, and Invention.* University of Chicago Press. [This is a delightful book, especially for technically oriented people like me. The author traces the many steps involved with the development of the steam engine and its applications, and he explains the workings of each improvement. He shows how the science and technology of the time interacted strongly for the first time and how white-collar thinkers and blue-collar artisans began to work together.]

Routman, Regie. 2000. *Conversations—Strategies for Teaching, Learning, and Evaluating.* Portsmouth, NH: Heinemann. [The author, a friend of mine, is an internationally recognized elementary school educator who has written several books on teaching and motivating students.]

Sobel, Dava. 1995. *Longitude—The True Story of a Lone Genius Who Solved the Greatest Scientific Problem of His Time.* New York: Penguin Books. [The author gives a vivid description of the work of John Harrison, who invented the chronometer (actually several of them) that enabled Captain Cook to accurately measure his ship's longitude over a six-month voyage. She also describes the irrational resistance from some scientists at the time and the malevolent sabotage he experienced at the hands of competitors and bureaucrats.]

--------------. 2000. *Galileo's Daughter.* New York: Penguin Books. [This is a delightful book that shows the human side of Galileo based on letters between him and one of his daughters, a convent nun who was supportive to him to the end.]

Stark, Rodney. 2005. *The Victory of Reason—How Christianity Led to Freedom, Capitalism, and Western Success*. New York: Random House. [The author provides some good historical material that shows that the commercial activity of the Italian city-states was an important part of the development of capitalism. The book is in part a rebuttal to Max Weber's assertion that capitalism developed primarily in northern Europe, driven by what he called the Protestant ethic.]

Teresi, Dick. 2002. *Lost Discoveries—The Ancient Roots of Modern Science, from the Babylonians to the Mayans*. New York: Simon & Schuster. [This is one of the first books I read when I decided to write my own book. The author does an excellent job of showing that innovations, both white-collar and blue-collar, have occurred over most of our history and in most of our civilizations. He avoids the trap of cultural superiority and gives appropriate credit to the innovative societies of the past. His examples are vivid and relevant.]

Usher, Abbott Payson. 1929. *The History of Mechanical Inventions*. New York: McGraw-Hill. [This book, referenced by William Rosen, goes into great detail to describe the development of some inventions of the Middle Ages, such as clocks and watches. He shows how these inventions were really many inventions, each building on previous experience. He strongly believes, like Conner, that many highly talented artisans and workers were behind most innovations and that Carlyle's claim that history is essentially the history of a few geniuses was totally at odds with the facts. He also rejects the idea that somehow society "produced" innovators, thus denigrating the efforts of individuals. I subscribe to his ideas and also believe that chance and luck, both good and bad, played an important role in the history of innovation.]

Yunus, Muhammad, 1999. *Banker to the Poor—Micro-lending and the Battle Against World Poverty*. New York: PublicAffairs™ of Perseus Books. [The author describes his efforts to bypass middlemen in the villages and rural areas of developing countries, and to provide short-term capital directly to industrious people who make things. His ideas on how to run micro-lending businesses are quite novel, and one hopes that this kind of activity will prove viable in the long run in more places in the world.]

Zakaria, Fareed, 2003. *The Future of Freedom—Illiberal Democracy at Home and Abroad*. New York: W. W. Norton & Co. [The author describes how the presence of elections does not guarantee functioning democratic institutions. He also claims that there are many institutions, both governmental and not, that are "too democratic" in the sense that the goals of some institutions within societies would be better met by having professional experts with a long-term viewpoint put in charge (such as judges in courts and financial experts in the US Federal Reserve Board) rather than be subject to the whims of politicians and the electorate, who tend to let short-term expediency dominate over long-term strategic thinking.]

Acknowledgments

This book was the outcome of a discussion group I was in several years ago. I had just retired from Trimble Navigation, where I had been working on incorporating advanced features into receivers for the Global Positioning System (GPS). The GPS had been the focus of my career for twenty years. Anticipating retirement, I formed a discussion group to discuss intellectual topics pertinent to life today. One of the topics that came up after reading Jared Diamond's *Guns, Germs, and Steel* was that of societies that seemed to be more successful than others in having inventors and natural philosophers. In chapter 13 of Diamond's book, he describes fourteen conditions that have been cited by various historians as being characteristic of innovative societies. In our discussions, we used our collective knowledge to expand the list. Keeping to the spirit of the original fourteen, we identified more than fifteen more, which we then proceeded to rank and evaluate among ourselves. The material in chapter 8 is based on the discussions we had. The discussion group consisted of the following members: Donald Barnett, a retired professor of English at DeAnza College; Michael Otten, a retired professor of sociology at San Jose State University; Patricia Helleboid, who obtained a solid classical education in France and teaches French; the late Howard Anawalt, who was a professor of law at Santa Clara University; and my wife Lois, a professional social worker. Long after they were ready to move on to other topics, I mulled over my notes and concluded that I could come up with a better list, one that satisfied a set of rigorous criteria like those we use in engineering. That started me on the delightful journey of research and synthesis that is described in this book. I want to give a special acknowledgment to my friends in this discussion group, who continued to give me support and acted as a sounding board for my ideas.

Later I met Tom Sawyer (yes, that's actually his name) and Hugh Treanor, both retired technical employees of IBM San Jose, and they gave me a lot of support and advice as I wrote the book. I owe them special thanks.

Several other friends gave me some useful inputs in specific areas. Tefcros Anthias reviewed my material on ancient Greece. Dominique and Claude Berger-Ceccaldi from Provence provided me with materials on Philippe de Girard from a number of French websites and contacted the librarians in Apt on my behalf. Alessandra Contenti, a professor of

English literature at the University of Rome and friend from graduate school, provided me with valuable material on medieval Italian history. Laurent Pfertzel, my neighbor and an entrepreneur, helped me with translating much of the technical French material on de Girard. Regie Routman, a well-known educator and writer, and Elliott Seif, an educational consultant, both gave me some excellent inputs on education and creativity. And Vladimir Vaganov, an expert in micro-mechanical systems, gave me valuable comments on the experience of the Soviet Union.

In addition, I want to thank a number of other friends who gave me support, encouragement, and advice throughout the writing of this book: Deanna Barnett, Anthony Goschalk, Ralf Gschwend, Len and Joan Kurzweil, Carolyn McDonald, Emerson and Cheryl Martlage, Evy and Joe Megerman, my cousin Ralph Pledger, Frank Routman, Thea Sawyer, Barry Slater, Ted Van Nahl, Beth White, Lynn Williams, Carla and Piero Sebastiani (of Florence, Italy), and Lynn and Don Yansen. Don, in particular, was an enthusiastic supporter.

I also want to thank the following: the staff at the Provence libraries in Apt and Lourmarin for their help and for allowing me to take copious notes from their resources; the National Gallery of the Marche in the Palazzo Ducale in Urbino, Italy, and the Cistercian Sénanque Abbey in Provence, France, for displaying plates of medieval construction machines; and the anonymous volunteers at Wikipedia, whose work enabled me to quickly gather specific information on a number of topics, an effort that just twenty years ago would have involved countless trips to the local libraries.

Finally, I want to thank my wife, Lois, who has been unbelievably patient and supportive to me—and who reviewed the book and helped me limit the amount of technical jargon involved—and my sons, Dan and Alex, who gave me unconditional support, advice, and encouragement.

Notes

CHAPTER 1: INTRODUCTION

No references

CHAPTER 2: INNOVATION IN EARLY SOCIETIES

1. Conner, 20–21.
2. Conner, 76–77.
3. Teresi, in numerous sections about Egypt on astronomy, mathematics, geology, and technology. The historical references in this section are from his book, unless otherwise noted.
4. Burns and Ralph, 37.
5. Teresi, 41.
6. Teresi, 332.
7. Exhibit at the de Young Fine Arts Museum of San Francisco, *Hatshepsut: From Queen to Pharaoh,* October 15, 2005 – February 15, 2006.
8. Burns and Ralph, 51.
9. Teresi, in numerous sections about Mesopotamia on astronomy, mathematics, geology, medicine, and technology. The historical references in this section are from his book, unless otherwise noted.
10. Diamond, 217.
11. Teresi, 233, 327.
12. Burns and Ralph, 72.
13. Burns and Ralph, 77.
14. Teresi, 233, 234, 329.
15. Burns and Ralph, 81.
16. McClellan and Dorn, 130.
17. Teresi, 351.

CHAPTER 3: A DIFFERENT KIND OF SOCIETY

1. The historical references in this section are from Cahill, unless otherwise noted.
2. Teresi, 17.
3. Cahill, 57.

4. Cahill, 55-60.
5. Cahill, 115.
6. Cahill, 59.
7. Cahill, 86, 129.
8. Cahill, 17.
9. Cahill, 17.
10. McClelland and Dorn, 71.
11. McClelland and Dorn, 77.
12. McClelland and Dorn, 79.
13. McClelland and Dorn, 83.
14. Burns and Ralph, 248.
15. Burns and Ralph, 249.
16. Burns and Ralph, 279.
17. Teresi, 247.
18. McClelland and Dorn, 89.
18. McClelland and Dorn, 91.
20. *Wikipedia*, "History of Rome" entry. See [http://en.wikipedia.org/wiki/Population_of_Rome#Sack_of_Rome_and_Counter-Reformation].

CHAPTER 4: THE MEDITERRANEAN LOSES ITS INNOVATIVE SPIRIT

1. Landes, 37–38.
2. Burns and Ralph, 365.
3. McClelland and Dorn, 99.
4. McClelland and Dorn, 117–140. The historical references in this section are from the books in this and the following reference, unless otherwise noted.
5. Teresi, 355–367.
6. McClelland and Dorn, 146.
7. McClelland and Dorn, 147.
8. Teresi, 354.
9. McClelland and Dorn, 148.
10. McClelland and Dorn, 125.
11. Teresi, 355.
12. Teresi, 245.

13. Needham, Appendix I.
14. McClelland and Dorn, 135.
15. McClelland and Dorn, 119.
16. McClelland and Dorn, 122.
17. Teresi, 72.
18. Teresi, 140.
19. Teresi, 250.
20. Teresi, 248.
21. Teresi, 82, referencing Kaplan, 56–57.
22. McClelland and Dorn, 105.
23. Rosen, xi.
24. Landes, 54 and chapter 24.
25. Landes, 45.
26. Landes, 53.
27. The National Gallery of the Marche (Galleria Nazionale delle Marche) in the Palazzo Ducale in Urbino, Italy.
28. Sénanque Abbey (Abbaye Notre-Dame de Sénanque) near Gordes in Provence, France.
29. Landes, 46–51.
30. Aczel, 106.
31. In this book, a "plutocracy" is defined as a societal form where a few powerful people share power as a group and exercise control over the rest of the society.
32. Acemoglu and Robinson, 130–134.
33. McClelland and Dorn, 181.

CHAPTER 5: INNOVATION GATHERS STEAM

1. Maddison, Angus, *Historical Statistics for the World Economy: 1–2003 AD*, © Angus Maddison. Here, personal well-being is associated with the gross domestic product (GDP) per capita as measured in Geary-Khamis dollars (GK$) referenced to the year 2000; the spreadsheet can be downloaded from the website [http://www.docstoc.com/docs/107349664/Historical-Statistics-for-the-World-Economy---GGDCnet]. *All GDP per capita figures cited in this book were obtained from this database.* See the bibliography reference *Groningen Growth and Development Center* for the meaning of Geary-Khamis dollars.

2. *Wikipedia*, "Republic of Venice" entry. See [http://en.wikipedia.org/wiki/Republic_of_Venice and http://it.wikipedia.org/wiki/Repubblica_di_Venezia].
3. Conner, subtitle.
4. *Wikipedia*, "Antonio de Ulloa" entry. See [http://en.wikipedia.org/wiki/Antonio_de_Ulloa].
5. Rosen, 125.
6. Rosen, 32–43.
7. Rosen, 39.

CHAPTER 6: INNOVATION TAKES OFF

1. This chapter draws heavily on the book by Rosen and its references. In this delightful account (at least for engineers like myself), the author makes a convincing case that Britain played the central role in the Industrial Revolution and in the establishment of invention as the source of societal prosperity.
2. Rosen, v.
3. The Malthus trap is the claim that population grows geometrically while the food supply can only grow linearly. Put another way, the population doubles every few years, while in each of those periods the food supply only increases by a fixed amount. If this proved true in practice, at some point the population growth would be limited by the lack of sufficient food, and the result would be widespread starvation.
4. This didn't happen without a human cost. Many workers lost their jobs to the factories. The owners of the factories were motivated to keep the factories going continually, regardless of the resulting disruption to the workers' lives.
5. *Wikipedia*, "Pearl Street Station" entry. See [http://en.wikipedia.org/wiki/Pearl_Street_Station].
6. *Wikipedia*, "History of the Telephone" entry. See [http://en.wikipedia.org/wiki/History_of_the_telephone].
7. Micklethwait and Wooldridge, 39–54.
8. Kay, 79.
9. *The Economist* (May 19, 2012), 13.
10. Wikipedia, "Bradford Parkinson" entry. See [http://en.wikipedia.org/wiki/Bradford_Parkinson].

11. Bradford Parkinson, *Cannon's Law*, Institute of Navigation, 50th Anniversary Special Session, 1995.
12. *Wikipedia*, "Shannon–Hartley theorem" entry. See [http://en.wikipedia.org/wiki/Claude_Shannon%27s_Law].
13. *Wikipedia*, "Discovery of cosmic microwave background radiation" entry. See [http://en.wikipedia.org/wiki/Discovery_of_cosmic_microwave_background_radiation].
14. *Wikipedia*, "Building 20" entry. See [http://en.wikipedia.org/wiki/Building_20].
15. Jonah Lehrer, "Groupthink—The Brainstorming Myth," *The New Yorker Magazine Annals of Ideas*, January 30, 2012. See [http://www.newyorker.com/reporting/2012/01/30/120130fa_fact_lehrer].

CHAPTER 7: INVENTION SPURS INVENTION

1. Diamond, 259.
2. Diamond, 239–241.
3. Diamond, 259.
4. Landes, 45–59.
5. Rosen, 66.
6. Rosen, 17.
7. Rosen, ix, 5.
8. Conner, subtitle.
9. *Wikipedia*, "Storia della Posta" entry. See [http://it.wikipedia.org/wiki/Storia_della_posta].
10. Kay, 69.

CHAPTER 8: THE CONDITIONS THAT SUPPORT INNOVATION—PREVIOUS WORK

1. Diamond, 249–251.
2. Diamond, 412–417.
3. Landes, 6–12, 15–20.

CHAPTER 9: THE CONDITIONS THAT SUPPORT INNOVATION—A NEW LOOK

1. Conner, 1–25.

2. Rosen, viii.
3. Usher, 12–14.

CHAPTER 10: ABSENCE OF FEAR

1. Rosen, 47.
2. Mernissi, 91.
3. Sobel, *Longitude*, 11–13.
4. Landes, 341–342.
5. Rosen, 68.
6. Landes, 40–59.
7. Stark, 57–61.

CHAPTER 11: ACCESS TO INFORMATION

1. *Wikipedia*, "Storia della Posta" entry. See [http://it.wikipedia.org/wiki/Storia_della_posta].
2. Menocal, 33–35.
3. Menocal, 188, 194–195.

CHAPTER 12: GUIDANCE IN THE USE OF INFORMATION

1. Agricola, xxvi.
2. Rosen, 72.
3. Stark, 107–109.
4. Routman, 463–470.

CHAPTER 13: SOCIETAL REWARDS FOR INNOVATION

1. Diamond, 259.
2. Landes, 45–59.
3. Bodanis, $E = mc^2$, 57–67.
4. Bodanis, *Passionate Minds*, 19–28.
5. Rosen, 13. Robert Boyle is famous for a number of important scientific discoveries, among them the formulation of Boyle's Law, which states that if the temperature of a gas in a closed container is held constant, changes in absolute pressure are inversely proportional to the volume.
6. Rosen, 10. Otto Gericke developed the famous Magdeburg hemispheres. The two hollow bronze hemispheres were placed

together, and the air was pumped out. Each hemisphere was attached to a team of fifteen horses, and the teams unsuccessfully tried to pull them apart, thus demonstrating the powerful forces associated with a vacuum.

7. Sobel, 53.
8. Rosen, 49.

CHAPTER 14: MULTIPLICITY OF OPPORTUNITIES

1. Diamond, 412.
2. Landes, 93–95.
3. Cahill, 86.
4. Stark, 107–118.
5. *Wikipedia*, "Michelangelo and the Medici" entry. See [http://en.wikipedia.org/wiki/Michelangelo_and_the_Medici].
6. *Wikipedia*, "Dutch Empire" entry. See [http://en.wikipedia.org/wiki/Dutch_Empire].

CHAPTER 15: THREE GREAT INNOVATORS

1. This account of Galileo draws heavily on Fermi and Bernardini (1961), *Galileo and the Scientific Revolution*, and Dava Sobel (1995), *Galileo's Daughter*.
2. Fermi and Bernardini, 41–47.
3. Fermi and Bernardini, 13–17.
4. This account of Philippe de Girard draws primarily on Bocquet (2003) and Rampal (1863).
5. When my wife and I arrived at Lourmarin, we found that the de Girard family home is now a school and museum. There is also a street named after him, the Avenue de Philippe de Girard. The local library is nearby, and we were allowed to take a couple of reference books that mentioned him with us to the house we were renting, where I took notes. On a later trip, Dominique and Claude Cicaldi, our friends and B&B proprietors in Rustrel, arranged for us to visit the library in Apt, where we were shown some biographical materials and an appeal to the parliament by his niece after his death. It was very exciting to be allowed to read books that were 150 years old and that vividly described de Girard's accomplishments and travails.
6. Michelin Green Guide, *Provence* (2002), 220.

7. Rampal, 14–18.
8. Rampal, 21.
9. At the library in Apt in Provence, we were privileged to read an appeal by the contessa in 1853 to both houses of parliament claiming that the original commitment by Napoleon should be honored and that she should be awarded the million francs, as Philippe's rightful heir. That claim was turned down by the attorney general of France in 1860, so the misfortunes of the de Girard family continued.
10. This account of Thomas Edison draws primarily on Dyer and Martin (1929).
11. The Henry Ford, "Thomas A. Edison and the Menlo Park Laboratory." See [http://www.hfmgv.org/exhibits/edison/default.asp].
12. Sobel (1995) describes the incredible persistence of John Harrison, his treatment at the hand of the Board of Longitude, and the different versions of his clocks, each quite successful and better than the previous one.

CHAPTER 16: INNOVATION AND GOVERNMENTS

1. Acemoglu and Robinson, 158.
2. Lester Thurow, "An Establishment or an Oligarchy?" *National Tax Journal*, vol. 42, no. 4 (December 1989): 405–11. See [http://ntj.tax.org/wwtax/ntjrec.nsf/10D6374119E9C6688525686C00686D9D/$FILE/v42n4405.pdf].
3. Virtually all countries have practices that violate the purest form of free-market conditions—namely, tariffs, subsidies, and other restrictions on trade. Democracies generally have fewer of them.
4. Jacobs, 23–24.
5. Kay, 18.

CHAPTER 17: INNOVATION IN TODAY'S SOCIETIES

1. Acemoglu and Robinson, 25–31.
2. Some commentators and authors have characterized the current Chinese government as a dictatorship. I strongly disagree, because dictatorships have the feature that the dictator decides if and when to step down and who will take his place. Dictators can readily remove enemies from, or add their henchmen to, the power structure

and the institutions supporting them. None of these apply to China, which has a National People's Congress of about three thousand members, a State Council of thirty-five members, and a Politburo of eight to nine members, all of which act to make the takeover by one person difficult. The presidents serve for ten years, and the premier and Politburo undergo orderly periodic changes. This is plutocratic behavior, not dictatorial, and the distinction is important.

3. Mark Elliot, "The Real China Model," *New York Times*, November 13, 2012.
4. *Wikipedia*, "Chinese Academy of Sciences" entry. See [http://en.wikipedia.org/wiki/Chinese_Academy_of_Sciences].
5. I don't know how much independence each branch has to select its own projects or its own members, if they can compete with each other, or how much inter-branch mobility there is for the members. However, judging by its centralized administration and the importance of control to the party, my guess is that there is little independence, competition, or mobility. If this is an accurate assessment, the academy will probably have limited success, as these features are essential for innovative activity to flourish.
6. Patents By Country, State, and Year—Utility Patents (December 2011). See [http://www.uspto.gov/web/offices/ac/ido/oeip/taf/cst_utl.pdf].
7. Shimon Peres, 1999–2000 season, Foothill College Celebrity Forum Speaker Series, Los Altos Hills, CA. See [http://www.foothill.edu/celebrity/past.php].
8. Zakaria, 96–101. He describes a number of "illiberal democracies" that have or have had elections, but which in practice do not have stable democratic institutions.

CHAPTER 18: APPLYING WHAT WE HAVE LEARNED

1. *Wikipedia*. See [http://wikipedia.org/].
2. Radjou, Prabhu, and Ahuja, 73.
3. d.school, "Hasso Plattner Institute of Design at Stanford." See [http://dschool.stanford.edu/].
4. Personal note: I have had a lot of experience in my own career with coming up with ideas that I thought were quite good, only to find out that they weren't very new after all. One time, I was on an airplane reading a technical description of liquid-crystal displays, or LCDs. One of their useful properties is that with a voltage applied,

they become opaque, while without the voltage, they are clear. I thought about this, and I realized that if one had glasses made from such a material, it would be possible to block the vision of one eye and not the other. The voltages could be switched rapidly so that one would be alternately seeing out of each eye. This meant that if you had a dual-television camera with a lens spacing the same as human eyes, it would be possible for one to "see" out of the left lens with the left eye, then see out of the right lens with the right eye. By having this dual camera switch back and forth rapidly between lenses and synchronizing the switching with these special glasses, the viewer would see the scene in three dimensions! It wouldn't require any change in the TV signal or the viewer's TV other than a synchronizing pulse to align the switching of the glasses with the TV camera. I was pretty impressed with myself for a few seconds until I realized that thousands of engineers had almost certainly already thought of this, given the widespread commercial interest in TV media and 3D. Sure enough, a few months later I read an advertisement by Toshiba describing televisions using this principle.

5. Dyer and Martin, Kindle book, chapter 2, loc. 214.
6. Kay, 121.
7. Percy Bysshe Shelley, "Ozymandias", 1818:

    I met a traveller from an antique land
    Who said: `Two vast and trunkless legs of stone
    Stand in the desert. Near them, on the sand,
    Half sunk, a shattered visage lies, whose frown,
    And wrinkled lip, and sneer of cold command,
    Tell that its sculptor well those passions read
    Which yet survive, stamped on these lifeless things,
    The hand that mocked them and the heart that fed.
    And on the pedestal these words appear—
    "My name is Ozymandias, king of kings:
    Look on my works, ye Mighty, and despair!"
    Nothing beside remains. Round the decay
    Of that colossal wreck, boundless and bare
    The lone and level sands stretch far away.'

8. Landes, 172.
9. Jaron Lanier, "Fixing the Digital Economy," *New York Times*, June 8, 2013.

Index

CPSIA information can be obtained at www.ICGtesting.com
Printed in the USA
LVOW10s1821020614

388160LV00014B/281/P

9 780741 497659